Just-in-Time Logistics

KEE-HUNG LAI and T.C.E. CHENG

GOWER

© Kee-hung Lai and T.C.E. Cheng 2009

All rights reserved. No part of this publication may be reproduced, stored in a retrieval system or transmitted in any form or by any means, electronic, mechanical, photocopying, recording or otherwise without the prior permission of the publisher.

Published by
Gower Publishing Limited
Wey Court East
Union Road
Farnham
Surrey
GU9 7PT
England

Gower Publishing Company
Suite 420
101 Cherry Street
Burlington
VT 05401-4405
USA

www.gowerpublishing.com

Kee-hung Lai and T.C.E. Cheng have asserted their moral right under the Copyright, Designs and Patents Act, 1988, to be identified as the authors of this work.

British Library Cataloguing in Publication Data
Lai, Kee-hung.
   Just-in-time logistics.
   1. Business logistics. 2. Just-in-time systems.
   I. Title II. Cheng, T. C. E. (T. C. Edwin)
   658.5-dc22

   ISBN: 978-0-566-08900-8

Library of Congress Cataloging-in-Publication Data
Lai, Kee-hung.
   Just-in-time logistics / by Kee-hung Lai and T.C.E Cheng.
      p. cm.
   Includes bibliographical references and index.
   ISBN 978-0-566-08900-8
   1. Business logistics. I. Cheng, T. C. E. (T. C. Edwin) II. Title.
   HD38.5.L35 2009
   658.7--dc22

2009002466

Printed and bound in Great Britain by
MPG Books Ltd, Bodmin, Cornwall.

Contents

List of Figures	vii
List of Tables	ix
Acknowledgements	xi
About the Authors	xi
Preface: Just-in-Time Logistics	vii

Chapter 1	**Introduction**	**1**
	The Value of JIT	2
	The Value of Logistics	4
	Significance of JIT for Logistics	5
	Layout of the Rest of the Book	7
	History and Development of JIT	9
Chapter 2	**Basics of JIT**	**9**
	Principles of a JIT Programme	12
	JIT in Services	16
	Elements of JIT Logistics	18
	The Goals of JIT	21
	JIT and Competitiveness	24
	Limitations of JIT	25
	The Rationale for Implementing JIT	27
	Prerequisites to JIT Implementation	30
	Summary	31
	History and Development of Logistics	33
Chapter 3	**Basics of Logistics**	**33**
	Scope of Logistics	35
	The Logistics System	35
	Inbound and Outbound Logistics	38
	The Core Logistics Elements	39
	Supply Chain Management	40
	The Goals of Logistics	43
	Logistics and Competitiveness	44
	Summary	45
	Customer Service	47
Chapter 4	**JIT Logistics**	**47**
	Logistics Service Quality	55
	Order Processing	58
	Inventory Management	78

	Transportation Management	96
	Summary	108
	Quality Awareness in Logistics	111
Chapter 5	**Quality Management Issues**	**111**
	The Need for Quality in Logistics	112
	Quality and Quality Costs	113
	Quality of Services	119
	Quality Issues in JIT	121
	Quality Management Tools	129
	Summary	134
	Organizational Factors for Implementation	135
Chapter 6	**Implementation Issues**	**135**
	Developing an Implementation Strategy	141
	An Operational Plan for Implementation	143
	Pilot Projects	145
	Dealing with Employee Resistance	146
	Adopting Enabling Technologies	147
	Diffusion of JIT Logistics to Partner Firms	150
	Data Collection and Measurement Systems	153
	Performance Measurement	155
	A Ten-step Approach for the Implementation of JIT Logistics	159
	Summary	165
Bibliography		*167*
Index		*179*

List of Figures

Figure 1.1	The value chain	5
Figure 3.1	The development of logistics	34
Figure 3.2	The logistics mix	37
Figure 4.1	Service quality model	53
Figure 4.2	Importance-performance matrix	58
Figure 4.3	The customer order cycle	59
Figure 4.4	A JIT order cycle	62
Figure 4.5	The bullwhip effect in the supply chain	83
Figure 4.6	Advanced planning and scheduling	95
Figure 5.1	PDCA cycle of improvement	124
Figure 5.2	Cause-and-effect diagram	133
Figure 6.1	A ten-step approach for the implementation of JIT logistics	160

List of Tables

Table 2.1	Example definitions of JIT manufacturing	13
Table 2.2	JIT practices in manufacturing	15
Table 3.1	Elements of JIT and supply chain management	42
Table 3.2	Comparison of lean supply chain, agility and RSC	43
Table 4.1	Characteristic practices of JIT purchasing	65
Table 4.2	Purchase resistance and marketer response to JIT	66
Table 4.3	Competitive priorities and purchasing	68
Table 4.4	Comparisons of market, relational and JIT relationships on characteristics of exchange	69
Table 4.5	Inventory types and their purposes	80
Table 4.6	Pitfalls of inventory management and their symptoms	82
Table 4.7	Product characteristics and their implications for transportation	98
Table 4.8	Transportation selection criteria and the implications for firms and service providers	100
Table 4.9	Information desired for better transportation services	101
Table 5.1	Quality improvement system summary	115
Table 6.1	Common problems associated with implementing JIT logistics and the strategies to overcome them	142
Table 6.2	Operational success factors for the implementation of JIT logistics	143
Table 6.3	Dimensions of institutional isomorphism on the adoption of JIT logistics in the supply chain	154
Table 6.4	Performance measurements for JIT logistics in early to advanced stages	156
Table 6.5	SCOR performance measures	159

Acknowledgements

The authors wish to acknowledge Dr Christina Wong and Miss Ailie Tang for their able assistance in the development of this book.

About the Authors

Kee-hung Lai is an Associate Professor, specializing in logistics and operations management, in the Department of Logistics and Maritime Studies, The Hong Kong Polytechnic University. He received his Ph.D. in business from the same university. Dr Lai's research interests include business logistics, e-commerce and logistics, channel relationships and green supply chain management. He has co-authored two other logistics-related books, over 50 papers in such journals as *Journal of Business Logistics*, and provided consultancy and executive training services on logistics management to public and private organizations in Hong Kong and on the Chinese mainland.

T.C.E. Cheng is Chair Professor of Management in the Department of Logistics and Maritime Studies of The Hong Kong Polytechnic University. He is also the Siyuan Chair Professor of Nanjing University in China. He obtained his B.Sc.(Eng.) (First Class Honours) and M.Sc. degrees from the Universities of Hong Kong and Birmingham, respectively, and his Ph.D. and Sc.D. degrees from the University of Cambridge. Professor Cheng's research interests are in operations management and operations research. He has published over 400 papers in such journals as *Management Science*, *Operations Research* and *MIS Quarterly*, and four books. He received the Outstanding Young Engineer of the Year Award from the Institute of Industrial Engineers, USA, in 1992, and the Croucher Senior Research Fellowship (the top science award in Hong Kong) in 2001. He was named one of the 'most cited scientists' in All Fields, in Computer Science and in Engineering over the period 1998–2008 by the *ISI Web of Knowledge* in 2008. One of his papers published in the *European Journal of Operational Research* in 1989 was voted one of 30 influential articles published in that journal since its inception. According to the *ISI Web of Science*, he has attained an *h*-index of 22 (that is, having produced 22 publications each attracting 22 or more citations). A paper in the *Asia-Pacific Journal of Operational Research* in 2008 ranked him as the most productive and the highest-cited researcher in Operations Research/Management Science in Asia.

Preface: Just-in-Time Logistics

KEE-HUNG LAI AND T.C.E. CHENG

The enduring repercussions of the Asian financial crisis in 1997, the worsening global economy following the burst of the dotcom bubbles in 2001, the financial tsunami in 2008 and the incessant rise in customer demand for better services have all contributed to shrinking profit margins for all businesses in the world. To cope with these challenges, many firms are looking for ways to strengthen and preserve their market positions. Logistics is becoming increasingly popular as a competitive weapon, among others, for firms to gain advantage in cost and services. These competitive strengths are achieved through an emphasis on the seven rights (7Rs) principle in logistics for time and place utility creation, that is, the ability to deliver the right amount of the right product at the right place at the right time in the right condition at the right price with the right information to satisfy customers fully. The 7Rs principle in logistics is congruent with the thinking of Porter (1985) on competition, where firms can enjoy competitive advantage through cost reduction or service differentiation or a combination of both in a niche market. With enhanced ability to reduce cost and improve services through efficient and cost-effective logistics management, firms are better positioned to compete in the turbulent business world of late.

Just-in-Time (JIT) is a pertinent concept for firms to execute the 7Rs principle in logistics and achieve the dual cost and service advantages. Originating from the manufacturing sector, the main thrust of the JIT concept is on the elimination of waste, that is, anything that adds no value, in the manufacturing process. This book extends the JIT concept in manufacturing to business logistics, an area that has been observed to account for more than 30 per cent of sales revenue for some firms. Different from manufacturing logistics, which is primarily concerned with the management of material flows to support a production line, business logistics concerns mainly with move-store activities with a major focus on the physical flows of finished products across organizational boundaries and to the end customers.

Application of JIT in business logistics is far more than waste reduction via ordering in small lot sizes and more frequent deliveries in support of a production line. It deals with inter-organizational relationship management and performance measurement for need identification and service improvement, as well as the proper utilization of people and facilities for the creation of customer value in move-store activities. Generally, these move-store activities in business logistics can be categorized into four core elements, namely customer services, order processing, inventory management and transportation, and a host of supporting elements including packaging, labelling, warehousing, materials handling and so on.

The objectives of this book are twofold: 1) to present an overview and introduction to JIT in business logistics, and 2) to provide insights into how improved logistics performance in terms of cost and service can be achieved through practising the concepts introduced in this book. To these ends, this book examines different types of waste in business logistics and discusses ways to reduce waste for improved efficiency. It also explores the JIT concept from a service perspective by discussing how it can help firms to improve service so as to satisfy the various parties involved in the move-store activities. As an introductory text on this subject, much of this book centres on introducing how to integrate JIT with business logistics; and how the JIT concept and its associated supporting management tools can be applied to achieve cost and service advantages with a primary focus on the four core elements of business logistics. Implementation and performance management issues of JIT in business logistics are covered as well. Furthermore, this book is also designed as a companion of an earlier book *Just-in-Time Manufacturing: An Introduction* by T.C.E. Cheng and S. Podolsky, providing practitioners with references on the application of the JIT concept in logistics and operations management.

This book is organized into six chapters. Chapters One to Three cover the key concepts and developments of JIT and business logistics, and provide an overview of JIT application in business logistics. Chapter Four explores how the JIT concept can be applied to improve cost and service performance in each of the four core elements of business logistics mentioned earlier. Chapters Five and Six discuss the various management issues of the JIT concept in business logistics pertaining to quality management, implementation and performance measurement. A ten-step approach for JIT logistics implementation is provided to conclude this book.

CHAPTER 1

Introduction

In face of the challenges of global competition, business firms are concentrating more on the needs of customers and seeking ways to reduce costs, improve quality and meet the ever-rising expectation of their customers. To these ends, many of them have identified logistics as an area to build cost and service advantages. On the other hand, the Just-in-Time (JIT) management approach, which has long been proven effective in the manufacturing sector in increasing quality, productivity and efficiency, improving communication and decreasing costs and waste, might enhance the chances of firms to achieve cost and service advantages through logistics. However, the potential of JIT has not been widely recognized in logistics as compared to in manufacturing. Similar to manufacturing, logistics employs processes that add value to the basic inputs used to create the end product. As the focus of JIT is on business processes, not products, the management principles of JIT can be replicated and applied in logistics. This book sets out to explore the possibilities of employing JIT to manage logistics activities, and provide an introduction to the application of JIT in the major areas of business logistics, which mainly deals with inter-organizational move-store activities. These move-store activities in business logistics can, in general, be categorized into four core elements, namely 1) customer service; 2) order processing; 3) inventory management; and 4) transportation management, and a number of supporting elements including materials handling, packaging, purchasing, warehousing and so on. This book concentrates on illustrating how the JIT principles can be applied in business logistics with a focus on the four core elements. 'Logistics', 'business logistics' and 'logistics management' are used interchangeably in the text to collectively represent inter-organizational move-store activities.

The expanding global competition, emerging new technologies and improved communications have increased customers' expectation of full satisfaction with the products and services that they purchase. These changes have, in recent years, brought to many manufacturing and service firms the challenges of improving the satisfaction of their customers and the quality of their products and services. Faced with these challenges, business firms worldwide are prompted to look for ways to reduce costs, improve quality and meet the ever-escalating demands of their customers. One successful solution has been the adoption of JIT manufacturing systems, which involve many functional areas of a firm such as manufacturing, engineering, marketing and purchasing, among others.

In the past decades, JIT has been primarily applied to manufacturing. Its obvious application with measurable outcomes in manufacturing has made JIT relatively easy to employ in a manufacturing environment. Although JIT has achieved a strong foothold in manufacturing, its application in business logistics is relatively recent for many firms. Yet, many developed and developing economies are experiencing a rapidly growing service base. For instance, in Hong Kong, logistics has evolved as one of the pillar business sectors. Logistics activities and the associated import/export trade account for more than 20 per cent of the GDP of the economy. Increased growth and competition in the logistical flows of products and services are likely to lead firms to embrace the practices of JIT in order to be cost-effective. The shift from a production-oriented to a service-oriented economy has led to a surge in firms' awareness of the potential of logistics to gain cost and service advantages. Many of them, particularly those in the retailing and transportation logistics sectors, have been challenged to sharpen their focus on customer satisfaction and the quality of their move-store activities. In response, they have been examining ways to satisfy the evolving customer expectation in a cost-effective manner. In such service settings as the physical distribution of manufactured goods with repetitive operations, in high volumes, and with tangible items, the implementation of JIT can help firms to streamline their move-store activities with less cost and greater effectiveness.

Indeed, JIT is also valuable for improving the performance of activities in service contexts. Among others, business logistics is an area where the implementation of JIT can help attain its performance objectives, that is, cost reduction and service improvement. Deliveries of pizzas, express mail and fast food are some examples of services that can benefit from the implementation of JIT for cost and service improvements. A growing number of studies on the subject have been published since the 1990s and the adoption of JIT in service settings is fast becoming an elusive area for both academics and practitioners to explore. Here in this book, we attempt to provide an introduction of JIT application in a service-based manufacturing context, that is, business logistics, an important but under-explored area in which the JIT management principles can be applied to further enhance business performance. After the burst of the dotcom bubbles in 2001, many firms have woken up and reverted to recognize the importance of offline business processes in serving their customers. Among others, business logistics is attracting the greatest attention as an area for firms to achieve cost and service advantages. In this regard, we target this book on providing a timely and useful contribution on this subject.

This book has two broad objectives, namely: 1) to provide an overview and an introduction of JIT logistics; and 2) to provide managerial insights on how to achieve improved logistics performance in terms of cost and service enhancements through practising of the concepts introduced in this book. The first objective is to familiarize the reader with the overall JIT concept and its application in business logistics, as well as the factors necessary for its effective implementation. The second objective is to illustrate how the concepts introduced in this book can help a firm improve performance in business logistics from both the cost and service perspectives. A discussion of the quality, implementation and performance measurement issues related to the application of JIT in business logistics will be presented too.

The Value of JIT

JIT is a management approach, which originated in Japan in the 1950s. It was subsequently adopted by Toyota and many Japanese manufacturing establishments with considerable success in raising productivity by eliminating waste (Kaneko and Nojiri 2008). Since its wide application in manufacturing in the 1970s, JIT has been widely regarded as an operations management approach designed for manufacturing firms to improve performance through waste reduction. According to Chase et al (2006), waste in Japan, as defined by Toyota's Fujio Cho, is 'anything other than the minimum amount of equipment, materials, parts, and workers (working time) which are absolutely essential to production'. The management philosophy underlying JIT is to continuously search for ways to make processes more efficient with the ultimate goal of producing goods or services without incurring any waste.

The first to embrace the JIT management approach were the production plants in Toyota. JIT gained widespread recognition during the 1973 oil embargo and wide adoption in many other organizations later. The industry-wide diffusion of JIT was, to a large extent, due to the oil embargo and the increasing shortage of other natural resources. To stay competitive, firms need to look for ways to reduce waste in their business processes. To cope with the difficult economic challenges, Toyota managed to survive by adopting an innovative management approach, that is, JIT, which was vastly different from what was characteristic of the time, which focused on the integration of people, plants and systems to reduce waste in its manufacturing processes.

JIT is an integrated, problem-solving management approach aimed at improving quality and facilitating timeliness in supply, production and distribution (Davy et al. 1992). Toyota believed that the only way for JIT to be successful is to have every individual within the organization involved and committed to it, if the resources and processes are fully utilized for maximum output and efficiency, and if the product and service offerings are delivered to satisfy market requirements without delay. Even three decades later in the twenty-first century, many firms are still struggling with the JIT management approach. JIT has gained considerable interest because it allows a firm to deliver high-quality products/services with reduced waste and increased productivity.

The implementation of the JIT management approach requires a body of knowledge, encompassing a comprehensive set of management principles and toolkits. These principles and toolkits are to be introduced in later parts of this book. Generally, it is accepted that the implementation of JIT can lead to improved firm performance. For example, in a study on the financial impact of JIT adoption, Kinney and Wempe (2002) found that JIT adopters outperform non-adopters in asset turnover and profit margins. The reason behind this is that the ability of the adopters of JIT to turn their asset is increased with improved product quality, greater responsiveness to customer demand because of shorter lead times and greater product line variety. These performance dimensions are underpinned by the philosophical elements of the JIT management approach on waste reduction and system flexibility for the performance of business processes.

Kinney and Wempe (2002) suggested further that firms practising JIT are associated with increased profit margins as the waste reduction emphasis of JIT helps reveal activities that add no value. Generally, these activities and their related costs are either hidden by excessive buffer inventories, or are ignored because holding buffer inventories is a convenient solution to such problems as failure of production lines or other systems. With the implementation of JIT, excessive inventories are no longer allowed to mitigate these problems and the adopters of JIT are more inclined to develop cost-saving solutions, thereby increasing profit margins. Another study also found a positive relationship between the level of JIT implementation in US manufacturing firms and their performance improvements (Fullerton and McWatters 2002).

The Value of Logistics

Logistics refers to all the move-store activities from the point of raw materials acquisition to the point of final consumption. Its core elements include customer service, order processing, inventory management and transportation (Ballou 2004).

- *Customer services* relates to the quality with which the flow of goods and services is managed. It is about the creation of time and place utility in the seven rights (7Rs), that is, the ability to deliver the *right* product to the *right* customer at the *right* place, in the *right* condition and *right* quantity at the *right* time, at the *right* (lowest possible) costs.
- *Order processing* involves all the activities in the order cycle, including collecting, checking, entering and transmitting order information. It is the means by which firms in the logistics processes exchange order information. The information collected will provide useful data for market analysis, financial planning, production scheduling and logistics operations.
- *Inventory management* is about managing appropriate inventory levels to serve the demand in a supply chain.
- *Transportation* is concerned with the ways in which physical items, for example, materials, components and finished products, are transferred between different parties, for example, raw materials suppliers, distributors, retailers and end customers, in a supply chain.

Effective and efficient logistics management is key to the success of business firms. For instance, the role of logistics plays an important role in such retail chains as Wal-Mart and 7-Eleven. They sell identical commodity products, for example, Coca-Cola, Campbell Soup and Kleenex, as other household retailers do. The core competence of these retail chains is logistics management, not product design and innovation, which help them outperform their competitors and attain sustainable growth. Nevertheless, poor logistics management can result in higher logistics costs, for example, inventory, transportation, order processing, whereby firms handling the processes will suffer from the higher logistical costs, and consequently lower profitability and reduced competitiveness. Viewed from this perspective, logistics management is not only a 'good-to-have' business strategy but a 'must' to sustain the growth of a firm or even a supply chain in the long run.

The value chain concept of Porter (1985) provides further insights on how logistics can contribute to the cost and service advantage of firms. The value chain, depicted in Figure 1.1, illustrates the activities that a firm must perform in order to provide benefits to customers. Primary activities in the value chain include those involved in the ongoing production, marketing, delivery and servicing of the product or service. There are support activities including such primary tasks as purchase inputs, technology, human resources and overall infrastructure needed to support the primary activities. It is important to note that two of the five primary activities are related to logistics: supplying raw materials, component parts and related services into the production line (inbound logistics), and managing the flow of finished goods from the end of the production line to the customer (outbound logistics).

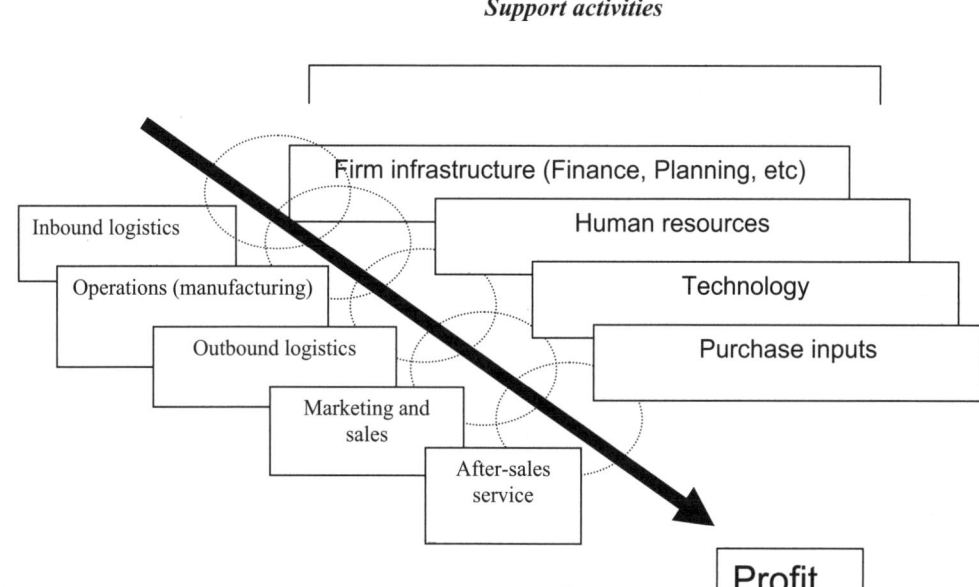

Figure 1.1 The value chain

On the other hand, past research has indicated that logistics influences a manufacturer's ability to satisfy customers and overall performance (Tracey 1998). It is important for firms to develop logistics capabilities to attain cost and service advantages (Lai 2004). Similarly, another study has found that logistics service performance engenders customer satisfaction, which has links with customer loyalty and market share (Stank et al. 2003).

Significance of JIT for Logistics

The goal of JIT in reducing waste and improving services is relevant and applicable to business logistics. Similar to manufacturing, JIT can be embraced as an operating management approach designed to eliminate waste. In business logistics, waste can be defined as anything other than the minimum amount of equipment, space and workers' time, which are absolutely essential to add value to the product or service. As logistics involves move-store activities in the supply chain, firms can embrace the philosophy of JIT to identify waste and improve service in the processes, for example, to plan the manpower and facilities requirements to meet the distribution needs, to reduce product introduction time by responsive delivery, to improve logistics service quality by forging supplier and customer partnerships and so on. In sum, there are many areas where the philosophical emphases of JIT on waste reduction and service improvement can be applied to improve logistics performance in the four core elements of business logistics, that is, customer services, order processing, inventory management and transportation.

WASTE REDUCTION

The primary objective of JIT practices is to eliminate waste. Wastes in JIT are not limited to tangible items such as excessive inventories and defective items, but also intangible items such as under-utilized manpower and facilities that have better use elsewhere. Waste can be generated in the different core elements of business logistics. One may wonder what kind of waste will be present in logistics activities. To deliver good customer service, flows of product/services must be managed well to meet customer requirements. Any activities that occupy time and consume resources in the product/service flows, but do not add any time and place utilities to the flows or the involved parties, for example, final customers, in the logistics processes can be considered as 'waste'. For example, delayed product delivery due to backorder by suppliers is a waste because the extra time needed to deliver the required items is of no value to the customers. In JIT logistics, all the activities that take up motion time, for example, order picking, shipment marshalling, transporting and so on, need to be managed efficiently and the performance outcomes need to conform to what are expected by the customers. The aim is to meet customer service requirements at the lowest possible cost.

Delay in information flow between suppliers and customers, for example, due to paperwork, is another example of waste because the administrative procedures may extend the order cycle time and result in higher safety stock against unforeseen events in the lengthened ordering process. Under JIT, the order processing procedures need to be simple and responsive. To these ends, the Continuous Improvement (CI) in the procedures involved in order cycles and the adoption of enabling technologies such as a Logistics Information System (LIS), often in the form of Electronic Data Interchange (EDI) or Value-Added Network (VAN), or the Internet are desirable so that different parties in the logistics processes can gain access to the needed information for decision making, therefore meeting the market requirements responsively.

Carrying large volume of inventories due to price discounts by suppliers or to prevent possible stock-outs are wastes in inventory management. These wastes often incur extra cash to finance, plus the manpower and physical space to handle the excess inventories. Under a JIT environment, raw materials, work-in-process (WIP) items and finished products are available in the exact quantities only when needed. Billesbach and Hayen (1994) argued that the adoption of JIT will lead to significant improvement in inventory efficiencies. This can be achieved through the elimination of unnecessary inventories, thereby shifting the resources to other revenue-generating activities.

Idle capacities for transportation facilities are wastes as well. For instance, queues of trucks for loading and unloading can be avoided if there is a proper loading schedule in the loading bay. There are also chances for trucks to maximize the carriage capacity if the routes are carefully planned such that trucks can be dispatched to different suppliers to gather partial shipments so that the transportation costs are reduced among suppliers, intermediaries and firms (Schnedider and Leatherman 1992).

SERVICE IMPROVEMENT

Quality of service in logistics is concerned with the achievement of the 7Rs. The role of JIT in service improvement is to help firms better understand their customers' requirements and their capabilities to satisfy those requirements. For instance, the practice of JIT enhances logistics customer services by ensuring the availability of goods to meet demand requirements, thereby creating time, place and possession utilities for the customers.

JIT also contributes to promoting close working relationships with suppliers to ensure the quality and dependability of supply. The pull-based nature of JIT requires a series of management practices. It will lead firms to continuously look for ways to improve their logistics activities with supplier firms to meet the JIT requirements. Some viable methods include supplier development and relationship management, which are to be discussed in later parts of this book, with just a few or even one supplier. This will assist in the creation of a more efficient firm in terms of inventory and materials, timeliness of deliveries and reassurances that the needed products/services will be available when required. On the customer side, JIT will assist the firm in focusing on what is demanded from customers and required of the offerings. The practice of JIT is consistent with the fundamental purpose of the firm to deliver products to satisfy customer wants. Developing a business process with an emphasis on JIT, which delivers quality products/services, will ensure the viability of the firm.

On the other hand, achieving quality products/services should not be carried out to the point where it does not pay off for the firm. Therefore, the emphasis should be placed on developing business processes that aim for zero defects. This may seem like an unrealistic goal; however, it is much less costly to the firm in the long run as it eliminates redundant functions such as inspection, re-work and complaint handling for defective products/services.

Layout of the Rest of the Book

In this chapter we briefly discussed what JIT and logistics refer to and how the implementation of JIT can help firms attain the dual objectives of waste reduction and service improvement in logistics. Chapters Two and Three will respectively discuss the evolution and development of JIT and logistics, and how they are related to enhance the competitiveness of a firm. Chapter Four will explore the roles of JIT in logistics, and discuss in detail how the principles of JIT can be applied in the four core elements of logistics. In Chapters Five and Six, the focus is on the operations issues of JIT logistics, including quality management, implementation and performance measurement. A ten-step approach for the implementation of JIT logistics is presented at the end of this book with a view to providing a roadmap for firms to embark on their JIT logistics journey.

CHAPTER 2
Basics of JIT

History and Development of JIT

JIT, a management approach originating in Japan in the 1950s, has been widely adopted by Japanese manufacturers since the1970s. Elimination of waste is the cornerstone of the JIT management approach. It is acknowledged that the implementation of JIT is one of the major factors contributing to the success achieved in the international competitiveness of Japanese manufacturers in the past decades (Wu 2003). The early development of JIT was initiated by Taiichi Ohno in Toyota's manufacturing plants in an attempt to meet customer demands precisely with minimum delays (Cheng and Podolsky 1996). Ohno (1982), the originator of JIT, defined JIT as making available the right part at precisely the right time, and in the right quantity, to go into assembly.

Owing to its relatively small geographical area and limited natural resources, Japan was compelled to find innovative ways to efficiently use its scarce resources. The Japanese have turned these disadvantages into advantages by successfully developing and implementing JIT production systems. They view the manufacturing process as a network of linked work centres that are optimally arranged such that each worker is able to finish their task and deliver it to the next worker exactly when it is needed. The ultimate goal is to completely eliminate all waiting time so that inventory investment can be minimized, production lead times can be shortened, demand changes can be quickly handled and quality problems can be uncovered and dealt with. JIT manufacturing strives to eliminate waste and reduce inventory by simplifying the production process (Schonberger 1986).

JIT gained extensive support in the 1973 oil embargo crisis when shortage of natural resources was the worst ever. Toyota managed to survive the difficult times with the unique JIT management approach. The philosophy of JIT was brought to a high level of sophistication and formalized into a management system when Toyota sought to meet the precise demand of customers for different models and colours in its automobiles with minimum delay. JIT, according to Toyota, requires organization-wide involvement and commitment, arrangement of plants and processes to promote maximum output, as well as scheduling of quality and production programmes to improve product quality and production efficiency (Cheng and Podolsky 1996).

There are strong cultural aspects associated with the emergence of JIT in Japan. The development of JIT within the Toyota production plants did not occur independently of these strong cultural influences. The Japanese work ethic is one of these factors. The work ethic emerged shortly after World War II and was seen as an integral part of the Japanese economic success. It is the prime motivating factor behind the development of superior management techniques, which would become the best in the world. The Japanese work ethic involves the following concepts:

- Workers are highly motivated to look for constant improvement over the required standards that already exist. Although high standards are currently being met, there exist even higher standards to achieve.
- Companies focus on group effort that integrates talents, knowledge-sharing, problem-solving skills, ideas and the achievement of a common goal.
- Work itself takes precedence over leisure. It is not unusual for Japanese employees to work 14-hour days. This contrasts greatly with the emphasis in western nations on time available for leisure activities.
- Employees tend to remain in the same company throughout the course of their career span. This gives them the opportunity to improve their skills and abilities at a constant rate while offering numerous benefits to the company. These benefits manifest themselves in employee loyalty, low turnover costs and fulfilment of company goals.
- There exists a high degree of group consciousness and sense of equality among Japanese employees. The Japanese are a homogeneous race where individual differences are not exploited or celebrated.

The JIT philosophy began to attract significant attention from the West in the late 1970s under the label of the 'Kanban' system. This was rather misleading as Kanban is only one part of the total JIT system. Kanban refers to a manual production control system using two cards to issue production orders: a requisition Kanban is used to authorize the movement of standard containers between work stations, and a production Kanban authorizes the production of parts to fill a standard container at a work centre. JIT is the generic term now widely used to describe this overall approach to manufacturing and logistics.

In the early 1980s, a new concept known as 'zero inventories' was introduced to the American manufacturing industry. The zero inventory concept calls for the transportation of materials from outside suppliers directly to the WIP area, where the required value-added activities through the manufacturing operations are performed, followed by the shipping of the finished products out of the door, all at a reasonable rate of time. This zero inventory concept would save companies the costs of inspection, stocking, materials handling, inventory tracking, carrying the inventory and the risks of damage and obsolescence. The concept now formally termed 'JIT inventory' has evolved into a corporate philosophy that seeks to do the process right the first time and to eliminate any non-value-added (NVA) activities. Any amount of time that a part is delayed, moved or inspected is referred to as NVA time. It is time wasted because no value is created for the customer when the product is not being processed.

In 1991, Lance Dixon from Bose Corporation, an audio system manufacturer, claimed to elevate the JIT concept to the next level by formally including a supplier's representative on-site at the customer's premises and called the new concept 'JIT II'. (Sowinski and Orton 2001) These representatives, called 'inplants', will place orders to their own companies. They also assist the customers in new product development and manufacturing planning. The advantages of doing so are twofold. Customers enjoy having experts around so that they learn about the constraints, which they may not know, during the early stage of their product design phase. On the supplier's side, they will be able to strengthen their ties with the customer. The inplants will also ensure that the suppliers will be able to fill the customers' requests. In short, 'JIT II' extends JIT by suggesting an integrated approach to supply and manufacturing between supplier and customer firms.

JIT II exists when a supplier's employee is located in the purchasing department of a customer, replacing the salesman and buyer. This inplant representative is empowered to use the customer's purchase orders to place orders with the supplying firm, participates in all the design and production meetings involving the supplying firm's product area, and has complete access to the customer's facilities, personnel and data. This step moves towards Supply Chain Management (SCM), but must be extended back to the supplier's supplier.

Another example of the link between the two concepts is the use of information, traditionally in the form of Kanban, to signal the approval to make or move an item. Sony used a direct form of this methodology as part of their SCM with SOMO (sell one – make one). Information transfers are used as a form of Kanban to indicate that an item has been used or sold and it needs to be replaced.

Under the JIT philosophy, activities such as moving parts, waiting for parts, machine set-up and inspection are referred to as wastes. The key point in understanding JIT is that JIT is a continuously goal-oriented process to eliminate waste and improve productivity. Waste occurs when activities are performed that do not add value to the products/services created. These wastes can affect the ability of firms to reduce costs and improve customer responsiveness (Fullerton and McWatters 2001). Waste can be physically identified in many forms such as scrap, rework, equipment downtime, excess lead time, overproduction, lower space utilization and so on. In the US, many firms are adopting the JIT philosophy to reduce these wastes and a study has found that JIT adopters improve financial performance relative to non-adopters (Kinney and Wempe 2002).

It is easy to understand the idea of JIT management to eliminate waste and improve services. But in reality these concepts are difficult to implement because JIT management has a high degree of cultural aspects imbedded in its development. Heiko (1989) suggested several relevant Japanese cultural characteristics that may be related to JIT as follows:

- JIT management allows an organization to meet customer demand regardless of the level of demand. This is made possible through the use of a pull system of production. The Japanese cultural characteristic that most relates to the demand pull concept is their emphasis on 'customer orientation'. Satisfying consumer needs quickly and efficiently is a priority for most business organizations in Japan.
- The degree of time lapse between material arrivals, processing and assembly of the final product for consumers is minimized by the JIT production technique. Production lead time minimization is possibly the result of the Japanese cultural emphasis on speed and efficiency. This may be due to the overcrowded living conditions that are prevalent in Japanese cities.
- JIT allows a reduction in raw materials, WIP and finished goods inventories. This frees up a greater amount of space and time between operations within plants. The corresponding cultural characteristic is a great concern for space due to a very dense population.
- The JIT production technique uses containers for holding parts. This allows easy identification and monitoring of inventory levels. The use of designated containers

within the production process may be due to the emphasis placed upon the use of proper types of packaging that exist when goods are purchased by consumers.
- An element of JIT production requires that the plant be clean, that is, there should be no waste present that may hinder production. The Japanese are concerned about the cleanliness of their environment, which may be due to limited space. A clean and uncluttered environment may give the illusion of a greater area.
- JIT production involves the use of 'visible signals' to display the status of machinery. The corresponding cultural characteristic involves the use of many signs displaying various products. Another contributing factor to the use of visible signals is the high literacy rate among Japanese people as compared to other countries.

The differences that exist between the culture of the Japanese and those of the other countries have led to the belief that JIT only works in Japan. These cultural differences that contribute most to this belief include the Japanese work ethic and the role of the unions in the working environment of many western countries. As the implementation of JIT management requires negotiating changes in work organizations, unionized facilities will resist adopting the practices associated with JIT. Drucker (1987) discussed the problems of existing union work rules and job classifications in the implementation of JIT systems.

Typically, unions are influential in 'blue collar' organizations, which are prone to adopt JIT management. They tend to exert influence upon management in developing policies that are more favourable to labour. Therefore, issues such as increasing leisure time for labour would be contradictory to the Japanese work ethic. This may explain some of the beliefs that JIT is only compatible with Japanese firms, but not others. On the other hand, research has shown that the potential benefits of JIT in waste elimination and productivity improvement do not just happen (Vohurka and Lummus 2000). Before a firm can reap the benefits of JIT, they must accept JIT as an organizational philosophy. It may require the firm to change or modify its operating procedures, its production or service systems, and in most cases its organizational culture.

The claim that JIT cannot be effective in non-Japanese firms has not been substantiated as many organizations outside Japan have successfully implemented JIT. In particular, many firms in the US have achieved performance improvement as a result of implementing JIT (White, Pearson and Wilson 1999). It should be noted that in organizations where a union plays an active role in bargaining for employee concerns, it is beneficial to engage union involvement before modifying any job procedures and organizational cultures for JIT implementation.

Principles of a JIT Programme

The management philosophy behind JIT is to continuously search for ways to make processes more efficient with the ultimate goal of producing goods or services without incurring any waste. Waste, as discussed earlier, is 'anything other than the minimum amount of equipment, materials, parts, and workers (working time) which are absolutely essential for production' (Aquilano and Jacobs 2001). Zipkin (1991) suggested that JIT can be approached from two perspectives: pragmatic JIT and romantic JIT. Pragmatic JIT promotes inventory reduction, but not zero inventory with a focus on the concrete details of the production process. Romantic JIT emphasizes a dramatic action with an emphasis on zero inventory, taking the idea to the extreme. Nevertheless, there is a lack of a universally accepted definition of JIT and this leads to some confusion about what exactly JIT is (Mia 2000). Since the earliest publications discussing JIT manufacturing, there have been a variety of JIT definitions proposed, which are summarized in Table 2.1. As noted by Hallihan et al. (1997), there is little agreement on a clear definition of JIT. The existing work generally includes a broad-based production system that incorporates the manufacturing practices of 'efficient material flow, improved quality and increased employee involvement' (for example, White and Prybutok 2001).

In sum, JIT is based on the concept of delivering raw materials just when needed and manufacturing products just when needed. The focus of JIT effort is on minimizing waste in the manufacturing system. The philosophy underpinning waste identification and its elimination is the basis upon which the JIT management approach is built (Karlson and Awstrom 1996). This is accomplished by: 1) minimizing raw materials, work-in-progress

Table 2.1   Example definitions of JIT manufacturing

Authors	Definitions
Voss and Robinson (1987)	JIT may be viewed as a production methodology that aims to improve overall productivity through the elimination of waste and that leads to improved quality. In the manufacturing/assembly process, JIT provides for the cost-effective production and delivery of only the necessary quality parts, in the right quantity, at the right time and place, while using a minimum of facilities, equipment, materials and human resources. JIT is dependent on the balance between the stability of the user's scheduled requirements and the supplier's manufacturing flexibility. It is accomplished through the application of specific techniques that require total employee involvement and teamwork.
Graham (1988)	A management philosophy or toolbox of techniques, which is based on making a significant improvement in operating efficiency through reduced inventory levels, lead times and overheads.
Im and Lee (1989)	The JIT system is a concept or philosophy that employs as tools several production management practices such as set-up time reduction, cellular manufacturing, level production planning, preventive maintenance, multifunctional workers, quality circles, Kanban, JIT purchasing, and so on. Because of its very nature, each company must develop its own JIT system.
Turnbull et al. (1992)	Matching the market with the manufacturing system, eliminating waste in all forms.
Fielder et al. (1993)	JIT can be viewed from a number of different angles, including people (attitudes, motivation, education in philosophy in JIT, training in procedures) and engineering (layout, product design for manufacture, set-up reduction).
Sohal et al. (1993)	JIT is essentially a philosophy more than a series of techniques, the basic tenet of which is to minimize cost by restricting the commitment to expenditure in any form, including manufacturing or ordering materials, components and so on until the last possible moment.

and finished goods inventory; 2) exposing process inefficiencies and streamlining operations, for example poor set-up times, poor maintenance, poor scheduling; and 3) reducing the inefficiencies and streamlining operations (Schonberger 1982b). These three goals are achieved through a set of ten complementary practices (Davy et al. 1992):

1. focused factory;
2. reduce set-up times;
3. group technology;
4. total preventive maintenance;
5. multi-skilled employees;
6. uniform plant loading;
7. Kanban systems;
8. quality control;
9. quality circles;
10. JIT purchasing.

Practices such as focused factory, uniform plant loading, and work and part standardization aid in streamlining operations and reducing inventory. Group technology also helps to streamline operations and reduce inventory. Set-up reduction, mixed model scheduling and Kanbans are aimed at reducing in-process and finished goods inventory. JIT purchasing directly affects incoming inventory levels. Techniques such as self-inspection, poka-yoke error prevention and jidoka (quality at source) improve the in-process quality of products. Finally, all of these practices require a motivated and multi-skilled workforce. A multi-skilled workforce has been found to make better contributions to various aspects of JIT implementation such as preventive maintenance, quality circles, set-up time reduction, and other quality improvement suggestions. These JIT practices in manufacturing are summarized in Table 2.2.

Many studies have examined the core elements of JIT, such as set-up time reduction, small lot production, the use of Kanbans, level production scheduling and preventive maintenance (for example, Schonberger 1982a), as well as issues related to JIT implementation. These issues include the relationship of JIT to other manufacturing practices (for example, Lee and Ebrahimpour 1984), supplier and customer relation (for example, Sakakibara et al. 1993) and JIT implementation (for example, Mehra and Inman 1992). The impact of JIT on performance, particularly manufacturing performance, has also been the subject of a number of studies. These studies have consistently found the use of JIT to be consistent with gains in inventory (for example, Callen et al. 2000), quality (for example, Fullerton and McWatters 2001), process management (Dreyfus et al. 2004), and productivity performance (for example, White et al. 1999). Several studies have also found evidence of improved business performance associated with the use of JIT. Gains in both financial (for example, Huson and Nanda 1995) and market performance (for example, Germain et al. 1996) have been observed.

What makes JIT a superior philosophy is its ability to achieve the dual advantages of cost reduction and service improvement simultaneously. JIT emphasizes simplifying the manufacturing process in order to quickly detect problems and force taking immediate remedial actions (Fitzsimmons 2006; Hernandez 1989). Several researchers have recognized JIT as a system-wide approach to manufacturing that focuses on the timely

Table 2.2 JIT practices in manufacturing

JIT element	Definition
Focused factory	A production strategy that is based on corporate strategy. It centres on simplifying the organizational structure, reducing the numbers of products or processes and minimizing the complexities of physical constraints.
Reduced set-up times	Reduction of the time and costs involved in changing tooling and other aspects required in moving from producing one product to another. This reduces lot sizes and the need for buffer inventories.
Group technology	Collecting and organizing common concepts, principles, problems and tasks. It avoids unnecessary duplication through standardization. It includes sequencing similar parts through the same machines and creating manufacturing cells for processing.
Total productive maintenance	Rigorous, regularly scheduled preventive maintenance and machine replacement programmes. Operators are actively responsible for the maintenance of their machines.
Multi-skilled employees	Extended training of employees on operating several different machines and involving in various tasks.
Uniform plant loading	Reduction in the fluctuations of the daily workload through line balancing, level schedules, stable cycle rates and market-paced final assembly rates.
Kanban systems	A card or information system that is used to 'pull' the necessary parts into each operation as they are needed.
Quality control	Quality is established as the top priority of the production system. Involvement in quality effort required by all aspects of the organization. Implementation of statistical quality control methods is an integral part of establishing both process and product quality.
Quality circles	Small groups are formed from employees doing similar tasks. The groups are created to encourage employee participation in problem solving and decision making.
JIT purchasing	A supplier participation and partnership programme. Receiving just the right parts just when they are needed. Suppliers, lot sizes and paperwork are reduced.

delivery of quality products sought by the customer and the elimination of waste (Chase et al. 2006; Krajewski and Ritzman 2005; Schniederjans 1993). The implementation of a JIT system yields minimum inventories by having each part delivered when it is needed, where it is needed, and in the quantity and quality needed to produce the product. A JIT system enables companies to operate efficiently with the least amount of resources, and thus improves quality, reduces inventory levels and provides maximum motivation to solve problems as soon as they occur (Krajewski and Ritzman 2005; Schniederjans 1993). It is not only an approach to managing inventory, rather JIT contains a body of knowledge and encompasses a comprehensive set of management principles and toolkits.

One should note the fact that the JIT philosophy has often been referred to as 'big JIT', whereas the operational aspects regarding materials flow in JIT systems are titled 'small JIT' or 'jit' (Hurley and Whybark 1999). The manufacturing practices that were presented above fall within the realm of small JIT.

In short, the objective of JIT can be simply stated as 'produce the right item, at the right time, in the right quantities'. In manufacturing, JIT has been credited with yielding many holistic benefits. These benefits include reduced inventory levels, reduced investment in inventory, improved quality of incoming materials and producing consistent high-quality products. Some additional benefits of JIT that have been achieved in manufacturing firms are: improved operational efficiency, uniform workstation loads, standardized components, standardized work methods, cooperative relationships with suppliers, closer collaboration with customers and improved customer satisfaction. Despite the delayed application of JIT in services, for example, in logistics management, there have already been some success stories, which indicate that many of these JIT benefits that have been achieved in manufacturing can be replicated in services, although sometimes in a slightly different form.

JIT in Services

Over the years, the JIT management approach has made a transition from application in internal production functions of manufacturing, purchasing and design to application in external service functions (Green and Inman 2005). However, the spread of the JIT movement from the manufacturing sector to service industries has been rather slow in both practice and research. While JIT is a process-oriented waste elimination management approach, the principles are relevant and applicable to both manufacturing and service firms employing processes and systems to perform the tasks of production and delivery of products or services. The major works on JIT applications in the service sector have been discussion of Wal-Mart's and Proctor & Gamble's Quick Response Programme, and similar JIT programmes developed by Wal-Mart to ensure their stores are capable of holding items in stock, without holding too much stock, with no stock-outs, whilst ensuring a no over-stocking inventory policy (Hill 2000). This approach to inventory management has enabled Wal-Mart to produce a better service, still at the lowest possible price, whilst permitting Proctor & Gamble to significantly increase its business with Wal-Mart.

As JIT has won the hearts and minds of manufacturing practitioners, advocates suggest service operators will also benefit from JIT. It is believed that JIT is applicable to the service sector as both manufacturing and service firms employ processes to create their final products and services. Benson's (1986) paper 'JIT: Not Just for the Factory' marked the first attempt to demonstrate the use of JIT in the service industry. In his article, Benson introduced several early adopters of JIT in the service sector:

- McDonald's emphasized that only necessary items are kept in work areas. In order to ensure process quality, it 'industrialized' the service delivery system so that all workers would be able to provide the same eating experience around the world.
- British Airways implemented quality circles as a fundamental part of its strategy to implement new service practices.

- Federal Express, aiming to improve process performance, redesigned the work flow so that the air flight pattern changed from origin-to-destination to origin-to-hub, with the freight being forwarded by outbound planes to destinations.

It was until the mid-1990s that JIT started to grow in the service sector (Yasin et al. 2003). Researchers claim that service environments with repetitive operations, high transaction volumes and involving tangible items, such as fast-food services and express courier services, are likely to benefit more from JIT (Whitson 1997, Canel et al. 2000). Some even argue that the lack of WIP and finished goods inventories in the service environment creates an ideal setting for JIT since there are no tangible 'wastes' to be dealt with. Indeed, JIT is applicable to a variety of services. Health care and accounting services are two of the many examples. In the case of health care systems, JIT eliminates the need for a central distribution centre, encourages flexible materials management, trains nurses with a variety of skills, and redesigns nursing units so that they are patient focused. JIT is also valued in accounting since it removes redundant control points, reduces the number of employees involved and integrates new technology to automate work flow. There is also a study on the application of JIT to improve inventory and purchasing management in hotel operations (Barlow 2002). On the other hand, there are investigations of JIT applications in the selling function. The object of JIT-selling relates to the delivery of zero-defect products and services in exact quantities at the precise times and places desired by customers while minimizing all types of waste (Germain et al. 1994). According to Green et al. (2008), a JIT seller exhibits: 1) the ability to build value throughout the selling process based on organizational capabilities to deliver zero-defect quality, zero variance quantity, precise on-time delivery and the ability to minimize total waste and total cost throughout the production and marketing processes; and 2) the ability to develop single source, on-site relationships with customers.

Recent improvements in information technology have also enhanced the ability of services to benefit from JIT systems. For example, bar-code technology, point-of-sale (POS) systems and the applications of Radio Frequency Identification (RFID) have made it possible to create a JIT replenishment system between suppliers and retailers, which improves customer service. Wal-Mart utilizes the Wal-Mart Information System that incorporates a Retail Link (involving bar-coding technology, POS systems and EDI) and a satellite network to better manage its inventory, achieve better demand forecasts and build more collaborative buyer-supplier relationships. The introduction of the automated teller machine (ATM) has also had a huge impact on improving and expanding banking services, and improving customer service and customer satisfaction. Airlines and hotels use reservation systems and differential pricing to create uniform facility loads. Such successes have influenced many other service-oriented organizations, including government agencies, to adopt some aspects of JIT. The increasing number of implementation of many aspects of JIT in service businesses and the increasing number of JIT success stories will accelerate the adoption of JIT in the service sector. If the service environment is suitable for the adoption of JIT, so will business logistics. We will discuss how JIT can be applied to improve the logistics processes. Before going further, we will first examine the elements of JIT logistics.

Elements of JIT Logistics

JIT can be seen as a new way of thinking, planning and performing with respect to logistics. JIT logistics consists of several components or elements that must be integrated together to function in harmony to achieve the JIT goals. These elements essentially include human resources, the logistics networks and the production, distribution, marketing and accounting functions of a firm. Similar to the JIT manufacturing system in Toyota, these elements can be broadly grouped together as people involvement, logistics network and systems.

PEOPLE INVOLVEMENT

To attain success in JIT implementation, it is essential to gain support and agreement from all individuals involved in the process. A firm shares interests with many groups having a stake in it, and vice versa. Typically, business firms' stakeholders include the owners (shareholders), customers, employees, suppliers and the community. If firms are to be profitable over the long term, and if firms are to be considered to be socially responsible, they must consider the needs of all of their stakeholders (Berman et al. 1999). This stakeholder orientation suggests that management must be concerned with more than the traditional economic aspects of JIT. Such non-economic issues as ensuring the proper treatment of employees, respecting employees, treating other stakeholders fairly, protecting the environment, providing for good quality of life in the community and other corporate social responsibility issues should be just as important to firms in their decision making as cost reduction and efficiency concerns. Therefore, support and agreement from the following groups are essential in order to reduce the amount of time and effort involved in implementing JIT and minimize the likelihood of causing implementation problems:

- Shareholders and owners of the company – emphasis should be placed on the long-term realization of profit, and so short-term earnings should be reinvested into the company to finance the various changes and commitments necessary for JIT success. It should be made clear that most of the benefits associated with JIT will only be realized over the long run.
- Labour organizations – labour unions and members should be informed of the goals of JIT and made aware of how the new system will affect the employees' work practice. This is important in winning the union and workers' support to assist in the implementation and to remove potential problems and difficulties. Education of employees, from managers to frontline workers, is essential so that they know what JIT is, why JIT is implemented in the organization and how they can contribute to make JIT a success (Zhu et al. 1994). Failure to involve labour organizations will result in a lack of understanding of management motives and cause fears of job loss on the part of the labour. This can lead to implementation hurdles such as non-cooperation and resistance to change. Union support is vital in achieving the elimination of job classification to allow for multi-skilled workers and company-wide focus.
- Management support – management support is often stressed as one of the most important factors to implement JIT successfully within an organization (Yasin and Wafa 1996). This involves the support of management from all levels. It also requires

that management be prepared to set examples for the workers and initiate the process to change attitudes. Striving for CI is not only required of the employees in the work processes, but must also be inherent in management attitudes.
- Government support – government can lend support to companies wishing to implement JIT by financial incentives or training support. This can provide motivation for companies to become innovative as it bears some of the financial burden associated with the costs of implementing JIT. Government regulation is another aspect that firms should pay attention to. For instance, the US Custom's 24-hour rule, which requires the submission of cargo data prior to vessel loading, will make a direct impact on how practitioners manage their supply chains.

Organization theory suggests that people will be more compelled to work towards goals when they are included in the development of the goals (Lines 2004). To this end, JIT extols the ideas of involving employees at different levels in the organization. The introduction of quality circles and the concept of total people involvement are examples of the avenues available for attempting to maximize people involvement through the use of JIT. However, the introduction of changes in a firm has the potential of triggering reactive behaviours from the individuals who may be subjected to these modifications. JIT represents one of these changes and can cause significant organizational transformations. Although these changes may bring benefits to the firm, reactive behaviours such as resisting the change by working against organizational goals may develop. Involving people becomes increasingly important at this point. Communication, training and increasing the values of workers' job can help alleviate reactive behaviours.

Without employee involvement and effective communication, progress on implementing various elements of JIT is likely to be slow, and in many cases resisted by employees who lack awareness and understanding of what is going on in the organization and their role in the new system being implemented. The organization culture must be such that it respects humanity, promotes teamwork and offers opportunities for individual learning and sharing of ideas. It will also be beneficial to the organization to switch to a team-based structure from a hierarchical structure (Power and Sohal 1997).

LOGISTICS NETWORK

Operational procedures, production or service systems and even organizational culture need be modified in order to reap the full benefits of JIT (Yasin et al. 2003). Numerous changes occur about the logistics network which encompasses logistics network design, multi-function workers, demand pull, Kanbans, self-inspection and CI. Each of these will be explained separately with relation to how they tie into JIT logistics.

Logistics network design and multi-function workers – JIT targets to supply the product/service when it is needed, how it is needed and in the exact quantity it is needed. JIT systems tend to have repetitive processes and predictable material flows. Firms with JIT processes often focus on the elimination of waste, which is defined broadly as anything that does not add value to a product. JIT emphasizes smooth and continuous process flows, where neither the goods/services supplied nor the receiver of those goods/services ever has to wait for one another. These allow firms to minimize WIP and finished goods inventories, and identify process inefficiencies and bottlenecks so that they can be reduced or eliminated. Under JIT logistics, the logistics network is designed for maximum resource flexibility and is planned according to supplier and customer requirements rather than processes. This type of logistics network requires the use of 'multi-function workers', that is, the focus shifts towards training workers and providing them with the skills necessary to perform many tasks, rather one or two highly specialized tasks.

Demand pull logistics and Kanban – JIT is a 'pull' system, in which the status of a worker or work station dictates the actions of others. This differs from a 'push' system, in which products are provided based on the supplier's ability to provide them, regardless of end users' needs. JIT systems use the pull method because of the closer coordination between inventory levels and production/delivery needs. The concept of demand pull involves the use of demand for a given product/service to signal when production and delivery should take place. Use of demand pull allows a firm to deliver only what is required in the right quantity and at the right time and place. Kanban is a Japanese word meaning signal and is usually a card or tag accompanying products throughout a production plant. Indicated on the Kanbans is the name or serial number for product identification, the quantity, the required operation and the destination of where the part will travel to. The use of Kanbans assists in tying or linking different production processes together. Cost savings from JIT programmes are a result of reductions in inventory and associated holding costs (improved cash flow), releasing space for additional revenue generating activities, and the transfer of labour costs to the distributor.

Self-inspection and CI – the use of self-inspection by each employee is done to ensure that their work processes add value to products/services and are of high quality. Self-inspection allows mistakes and low quality work to be caught and corrected efficiently and at the place where the mistakes initially occur. In complement, the concept of CI involves a change in attitudes towards the overall effectiveness of an organization. CI is an integral part of the JIT concept and, to be effective, it must be adopted by each member of the organization, not only by those directly involved with the logistics processes. CI requires that with every goal and standard successfully met, these goals and standards should be increased but always in a range that is reasonable and achievable. This allows a firm to constantly improve upon its operations and products/services, and, ultimately, its customer satisfaction.

Given the nature of JIT, quality management will play a most important role in ensuring that the quality standards set for the logistics activities are achieved. JIT quality involves 'quality at the source' (Hay 1988). The emphasis of quality at the source is congruent with the achievement of 7Rs in logistics. Quality at the source contrasts greatly with the traditional 'after the fact' approach to quality or rectifying the defects after performing the tasks.

SYSTEMS

Systems refer to the technology and processes used to link, plan and coordinate the activities and resources used in the logistics networks of a firm. Two such systems are Materials Requirement Planning (MRP) and Enterprise Resources Planning (ERP).

MRP is a 'computer-based' method for managing the materials required to carry out a schedule. It is a 'top-down' or 'consolidation' approach to planning, that is, it involves the planning of lower level products within the product family such as component parts (Plenert 1999). Planning for MRP can be broken down essentially into two parts. These include a production plan, which is a broad plan indicating the available capacity and the manner in which it is to be allocated in the logistics network, and a master production schedule, which is a detailed plan of what products to produce in specific time frames.

ERP is a software system that can be used to provide information on financial resources available to carry out the plans of MPR. It also provides a firm with a single, uniform software platform and database that will facilitate transactions among the different functional areas within a firm, and in some cases, between partner firms in a supply chain. ERP solutions seek to integrate and streamline business processes and their associated information and work flows. What makes ERP appealing to organizations is its increasing capability to integrate with the most advanced electronic and mobile commerce technologies.

On the other hand, everyone in the organizational hierarchy should belong to a team. Workers should be organized into teams so that they are always working together to reduce all forms of waste, not just the waste incurred within their scope of work. Suppliers and key customers should also be involved in the teams so that all the players of the value chain are well aware of the JIT goals and strategies, and are working together to improve all the processes. If workers are organized into teams, teams' performance should precede individuals' performance. Management of various functional units will all participate in the decision-making processes to promote understandability of functional interdependence and interactions (Germain et al. 1996).

The Goals of JIT

JIT management can be applied to the logistics processes within any firm. Yasin and Wafa (1996) identified six potentially beneficial attributes of JIT that can increase organizational efficiency and effectiveness. JIT:

- tends to eliminate waste in production and material;
- improves communication internally (in the organization) and externally (between the organization and its customers and vendors);
- has the potential to reduce purchasing costs, which are a major factor for most organizations;
- is instrumental in reducing lead time, decreasing throughput time, improving production quality, increasing productivity and enhancing customer responsiveness;
- tends to foster organizational discipline and managerial involvement;
- tends to integrate different functional areas of the organization, especially to bridge the gap between production and accounting.

These attributes of JIT, if successfully applied in business logistics, should yield similar benefits to those found in manufacturing. For instance, waste in business logistics can take many forms, which include excessive inventory, overestimates of customer demand and extended cycle lead time. Elimination of waste can be achieved through streamlining the logistics process with proper logistics network planning, and reducing set-up time to adhere to the daily schedule made possible by a pull system, which extends from suppliers to manufacturers, and ultimately to customers, that is, the supply chain.

Porter (1985) suggested that a firm can achieve 'competitive advantage' by competing on the basis of cost, service or quality. These three elements are the distinguishing characteristics that differentiate the products/services of a firm from one another. JIT allows firms to filter out wastes in their logistics processes, improve upon quality and achieve customer satisfaction in an effective and efficient manner.

There are three main objectives of JIT, which are universal or homogenous in nature, that is, they can be applied and adapted to a diversity of organizations within industries that differ greatly from one another.

Increasing the organization's ability to compete with rival firms and stay competitive over the long run

Organizational competitiveness is enhanced through the use of JIT as it allows organizations to develop an optimal process for managing the logistics of their products/services. There are differences in the logistics processes between traditional and JIT-oriented firms.

The traditional firm is one that adheres to the well-practised forms of performing tasks. The JIT-oriented firm is one that can respond to changes within the environment and adapt its manufacturing processes to changes. Frequently, these types of firms are the first to develop or implement innovative methods for logistics. Thus, the JIT-oriented firm is one that is able to remain competitive through adaptation to environmental changes.

The JIT-oriented firm will have a well-integrated system of logistics that involves shared organizational values, coordinated flows in the logistics processes, people involvement and the opportunity to use potential skills. The differences that exist between the traditional and JIT-oriented firms involve operational and organizational characteristics.

The operational characteristics include set-up time, order size, inventory, capacity, transportation, lead time, defect rates and equipment trouble. It is typical for traditional firms to experience long set-up, transportation and lead times. Inventory, capacity and order size are likely to be large. In addition, defect rates and equipment trouble will be high for the traditional firm as well. In contrast, JIT-oriented firms will have short set-up, transportation and lead times. Inventory, floor space and order size will be small, and defects and equipment trouble tend to be low for these firms. The overall functioning of the logistics networks in JIT-oriented firms will be smoother and more efficient than those in conventional firms.

The organizational characteristics include the structure, orientation towards goals, communication, agreement, union focus, skill base, suppliers, and education and training. The structure of the JIT-oriented firm allows greater flexibility. The goal is towards total optimization of the whole company while avoiding departmental focus, which tends to work against the achievement of organization-wide goals. Communication within the JIT-oriented firm is open and there is not a long chain of command to follow. Also, agreement among members is trust-based as compared to contract-based. Union focus is company-based rather than skill-based. The skill base tends to be broad and flexible in contrast to narrow or highly specialized skills. The level of supplier involvement is narrowed down to include a selected few, and education and training play a significant role. These types of firms are more likely to invest more resources in training employees.

Increasing the degree of efficiency within the logistics process

Efficiency will concern itself with achieving higher levels of productivity while minimizing the associated costs for performing the logistics activities.

Reducing the level of wasted materials, time and effort involved in the logistics activities

Elimination of waste can significantly reduce the costs of production.

The above three universal objectives of JIT are applicable to any firm; however, there exist several other goals that may be specific to firms. In order for JIT management to work and be profitable, it must be fully adapted to the firm. Every firm is unique in its logistics processes and the goals it aims to achieve. In addition, every firm will be at different stages in its development. The goals of each firm are unique in terms of their priority and importance. The goals of JIT are useful in assisting the firm to define, direct and prepare for implementation. There exist short- and long-term goals, which include the following:

- Identifying and responding to consumer needs – this goal will assist the firm in focusing on what the customers need and what products/services need to be produced. The fundamental purpose of the firm is to produce products/services that its customers want, therefore, developing a process that delivers quality products/services to ensure the organization's viability.
- Aiming for the optimal quality-cost relationship – achieving quality should not be done to the point where it does not pay off for the firm. Therefore, emphasis should be placed on developing a process that aims for 100 per cent conformance. This may seem like an unrealistic goal; however, it is much less costly to the firm in the long run as it eliminates redundant functions such as inspection, rework and the production of defective products/services.
- Eliminating waste – this is anything that does not add value to the products/services.
- Aiming for the development of trusting relationships between the supply chain partners – for instance, relationships with just a few or even one supplier, if possible, should be focused upon. This will assist in the creation of a more efficient firm in terms of reductions in inventory and use of materials, enhancing the timeliness of deliveries and providing reassurance that resources will be available when required.
- Designing the logistics network for maximum efficiency and ease of management – this involves the use of machinery and labour that are absolutely essential to the logistics process.
- Adopting the Japanese work ethic of aiming for CI even though the firm is competitive, by continually striving to fulfil consumer demand.

Although many manufacturing and service firms have adopted the JIT management techniques, these firms are at the beginning stage and have not yet fully realized all the potential benefits. It has taken Toyota 10 years to perfect the JIT technique within its plants. Yasin and Wafa (1996) suggested that the JIT system must be willing to make strategic adjustments consistent with the demands of its environment. They believed that these long-term strategic adjustments are not feasible without short-term costs. They remarked that in most cases a system-wide strategic adjustment will not yield the desired results unless the subsystems, mainly the input, process, output and managerial subsystems, are modified to make feasible system-wide changes. The researchers reasoned that all the subsystems must actively contribute to and facilitate the incoming changes to ensure that the changes are successful. Physical facilities, for example, plant layouts, are also important according to the study. They must be adjusted, relationships with vendors and customers reviewed, and quality circles implemented. Therefore, JIT is a long-term process that cannot be implemented in a short period of time, nor can its rewards be realized overnight.

JIT and Competitiveness

A systematic and continuous pursuit of waste identification and elimination can lead to increased efficiency, improved productivity and enhanced competitiveness. JIT is expected to improve firm performance and competitiveness through an even production flow of small lot sizes integrating schedule stability, product quality, short set-up times, preventive maintenance and efficient process layout (for example, Chapman and Carter 1990). Generally, firms that work towards the elimination of waste in their manufacturing processes realize such benefits as lower raw materials stock and associated holding cost, reduced WIP, lower finished goods inventories, higher levels of product quality, increased flexibility and ability to meet customer demands, lower overall manufacturing costs and increased employee involvement (Canel et al. 2000). To a large extent, inferior competitiveness results from the existence of large amounts of waste and the reduction of these non-productive activities, that is, waste, would help to save time and allow more resources to be allocated to improving throughput and profitability (Cheng and Podolsky 1996). Such production improvements are assumed to bring both indirect and direct financial savings.

In theory, JIT improves profitability due to its impact on the two interdependent components of Return on Assets (ROA): asset turnover, which measures sales relative to investment; and return to sales (Kinney and Wempe 2002). JIT is expected to improve ROA in a number of ways. First, asset turnover should increase, as JIT frees up assets and capital. A smaller asset base increases ROA. Second, lower inventory levels reduce the asset base, improving asset turnover in the short term. Third, few buffer inventories necessitates the elimination of NVA activities (for example, dealing with defects and stock-outs) that have a negative impact on profit margins (Alles et al. 1995). As emphasized by Balakrishnan et al. (1996, pp. 185–186), these effects are not necessarily automatic, and can be offsetting, especially in the short term. For example, firms may be required to invest in additional training and capital expenditures to nurture a JIT environment. Training costs initially reduce profit margins, but are expected to improve long-run productivity. Capital expenditures increase the asset base and depreciation expenses in the short run, thus affecting both components of ROA.

There is also other published evidence that supports the positive impact of JIT on the competitiveness of firms. Inman and Mehra (1993) found a significant correlation between self-reported improvements in profitability and the adoption of JIT practices.

Callen et al. (2000) surveyed 100 Canadian plants from the auto parts and electronic industries, classifying the plants as either JIT or non-JIT. Their results show that JIT implementation leads to higher profit and contribution margins and lower variable costs. In another study, Callen et al. (2005) found that JIT-intensive plants are more profitable but less efficient than plants that are not JIT-intensive, after controlling for productivity measures, plant size and buffer stock. Their result suggests that despite the additional resources requirement for the implementation of JIT, JIT-intensive plants are still able to generate relatively higher profits than plants that are not JIT-intensive. Furthermore, Ward and Zhou (2006) confirmed that implementing JIT practices improves lead-time performance. In evaluating the JIT production practices of 46 Japanese manufacturing plants, Matsui (2007) found that JIT production systems contribute to improving competitive performance, and that efficient equipment layout has a strong impact on the competitive position of the manufacturing plant. In sum, there are several potential benefits from the implementation of JIT; in particular, JIT can help to:

- eliminate waste in production and materials (Cheng and Podolsky 1996);
- improve communications internally (within the organization) and externally (between the organization and its suppliers and customers) (Germain and Droge 1997);
- reduce purchasing costs that is a major cost to most organization (Dong et al. 2001);
- reduce lead time, decrease throughput time, improve production quality, increase productivity and enhance customer responsiveness (Vokurka and Lummus 2000);
- foster organizational discipline and managerial involvement (Yasin et al. 1997);
- integrate different functional areas of an organization (Gunasekaran 1999).

On the other hand, Fullerton and McWatters (2001) studied the correlation between the level of JIT practised and the production performances of 89 sampled firms. They evaluated production performance with respect to quality benefits, time-based benefits and employee flexibility. They found that:

- the more JIT practices are adopted by firms, less scrap and rework are anticipated, and the number of inspections decreases since in a perfect JIT world, inspections would not be necessary as there will be no defects;
- NVA activities including queue times, move times, machine downtimes and throughput times reduce with an increased level of JIT;
- employee flexibility is enhanced in high JIT adopters.

Limitations of JIT

Considerable attention has been paid to the benefits that can be derived from the use of JIT. However, in order to properly implement JIT in a firm, managers should be aware of the limitations and shortcomings of JIT, which may be applicable to their organizations. Several shortcomings have been identified as follows:

- The design of a JIT workflow is often such that there are no buffer stocks between different stages and a tight synchronization of work flow is emphasized. Workers are expected to work in a structured manner and identify any non-conformance

along the process. This may sometimes lead to a potentially difficult situation for the workers. On the one hand, they are voicing out the flaws for the sake of improving product/service quality; on the other hand, they are pinpointing their co-workers who are responsible for the flaws. It is not easy for workers to achieve this level of consciousness if the firm has a high social culture.
- Hall (1989) also claimed that the absence of buffer stocks makes JIT incapable of coping with a sudden surge in demand. The increase in demand will lapse because of the firm's inability to fill the extra orders. However, Germain et al. (1994) found that organizations faced with more volatile environments increasingly adopt JIT as a strategic response. They argued that JIT buffers firms from uncertainties because the improved overall customer service, quality and productivity enhance the appeal of a supplier to a buyer, and therefore this competitive edge can be used to obtain longer contracts, increase market share and profits, and pressure less competitive rivals.
- According to JIT, materials are transferred to the next stop only when needed. A tight transition between the vendors and their carriers is essential as any delay in the transportation process will hamper the production schedules. A well-planned transportation procedure is the first step to prevent shipment from delays. Nevertheless, such a procedure is still exposed to unexpected disruptions such as natural catastrophes, man-made accidents, labour disputes, equipment breakdowns and government regulations (McGillivray 2000). Ford Motor Company shut down five of its US factories shortly after the 9/11 terrorist attacks partly due to the late deliveries of Canadian parts (Aichlmayr 2001). To safeguard productions against disruptions, firms sometimes maintain marginal stocks to ensure continuous production.
- For more than a decade, JIT manufacturing has been touted as a prime way to keep costs down and assembly lines running smoothly. But because JIT requires plants to keep trimming inventories, even the smallest glitch in the supply chain can bring production to a standstill. JIT systems are designed so that parts and components arrive at factories just as they are needed for assembly, reducing and sometimes eliminating the need for warehousing expensive parts. Billions of dollars are saved on inventory, but when a plant making a critical part shuts down, there is little buffer stock from which to draw.
- Reliable suppliers seem to be at the heart of the problem. Most firms are still struggling to develop a critical mass of reliable, cost-effective, lead-time-stable suppliers that will allow them to operate their production lines on a JIT or demand-pull basis. It also appears that most companies continue to rely on forecasts (often from sales and marketing departments); only a handful have begun to provide suppliers with consistent real-time access to their ever-changing production schedules. Another difficulty is the removal of the 'human element' from the systems that generate requirements. Because computer algorithms are limited, there is still a need for good people with experience in detecting the ups and downs of the industry.
- One limitation of JIT is cultural differences; these differences play an important role when interacting with different firms in order to receive goods on time. Many organizations find it difficult to adopt new methods because of present cultures. In other words, the interacting people are set in their own ways and people are reluctant to make changes.
- Another limitation is the traditional way of having plenty of inventories on hand to cover ordering or product mistakes. Now with JIT firms are not able to have the

additional inventory that they are accustomed to, which creates a negative source of pressure on the participating individuals. Also the loss of autonomy on individuals and teams put more stress on firms because they only have a set amount of time to do certain tasks. The limit of trust between workers and their managers must be broken because firms cannot have any lack of commitment; trust must be achieved to have complete satisfaction of work with different types of equipment. Also a limitation is that firms can only achieve high production if every employee works equally as hard.
- JIT also involves a difference in the way in which managers think; employees' suggestions become the backbone of the organization. In addition, the elimination of costing and financial analysis may make managers feel they have lost control of the plant, because they now receive less data. The tendency of the stockholders to be short-term-profit-oriented causes managers to plan for short-term results rather than long-term successes. This short-term thinking may adversely affect the long-term development of a firm and must be curtailed if it is to compete in the global economy of the future.

The Rationale for Implementing JIT

The impetus for implementing JIT lies largely in attaining the productivity and quality standards that many Japanese firms have enjoyed. Economic conditions such as increased competition, fluctuations in the economy and consumer demand for high-quality products also play a role. Stiff competition has created an environment in which only the most effective and productive firms will survive. The organizations that are quick to apply innovative ideas to their manufacturing processes will have a competitive advantage over those that do not. These firms will be able to survive and be profitable over the long run.

The use of JIT is not adversely affected by fluctuations in the economy as production is readily flexible to meet variable consumer demand. The use of JIT is appropriate in both economic upswings and downturns as it can be adjusted rather painlessly to meet any consumer demand. This is possible as it operates on a pull system where demand acts as the impulse calling the production process into action. During economic booms firms and individuals have higher demands, and the production of products/services with JIT can be easily increased to meet these demands. Similarly, in economic downturns, production can be curtailed to meet lower levels of consumer demand.

Other reasons for adopting JIT are the potential cost savings associated with its use. Consider the profit formula: profit = selling price × sales volume − cost. This formula represents the components of profit. Most firms are unable single-handedly to influence the selling price of their products as the selling price is determined by the market forces of supply and demand and industry standards. Therefore, if firms wish to increase their profits, they must focus on increasing the sales volume and decreasing the cost. To increase sales volume requires better quality and delivery, while reducing costs calls for reducing any unnecessary operations and waste. JIT can help firms to improve the sales volume, reduce the cost component of their business processes and provide the opportunity to realize the increased profit.

The savings associated with cost reductions include reduced product recall, return and costs of lost sales and customer complaints. The logistics costs that can be reduced include costs for inspection, which are necessary for products/services of less than 100 per cent quality. With the achievement of the 100 per cent quality level, there is no re-working and testing required to improve quality. The cost of quality is also lowered by avoiding lost sales and costs associated with customer complaint handling. In addition, there will be less waste of time and effort in searching for flaws in the process that may be responsible for defective items.

JIT logistics focuses on adding value to products/services. This involves only performing the activities and processes necessary to deliver the products/services. To illustrate the concept of adding value, consider the following: customer order receipt, picking and packaging, and customer order delivery. Operations such as re-packaging, excessive handling and product returns do not add value to a product. They do not add to the time and place utilities of a product in any way, therefore these activities incurring unnecessary financial costs and effort invested in the logistics processes should be avoided.

Closely associated with the value-added concept is the idea of reducing waste. As discussed earlier, waste can be anything other than the minimum amount of equipment, material, parts and working time essential to deliver the products/services. There are seven forms of wastes: motion, waiting, time, overproduction, processing time, rejects, inventory and transport. Hallihan et al. (1997) provided practical examples and discussed the causes of these seven wastes in manufacturing, which are summarized in Table 2.3. Complete elimination of waste allows additional value-adding work to be performed in the time saved, yielding improvement in productivity.

1. The waste of motion. Motion study involves beliefs and practices developed through scientific management. The application of scientific management to JIT involves the idea that excess handling of materials and equipment to produce above demand require inefficient motions and involvement of employees. The motions required to move this excess of materials around the plant represent waste.
2. The waste of waiting time. This involves the length of time inventory in transit is idle and waiting to enter the next operations. Queuing time is largely the result of inefficient work flow and can cause uneven lot sizing.
3. The waste from overproduction. JIT manufacturing allows a firm to produce only what is needed, operating on the demand pull concept. Therefore, in many plants that do not use the demand pull concept, overproduction will occur. The wastes caused by overproduction are employee effort and time in producing products that are not required.
4. The waste from processing times. This includes the processing of parts that affect the final or finished product. These parts may or may not be a necessary step in the completion of the product. They also may not contribute to the value of the product.
5. The waste from rejects. The use of inspection after the product is made or partially completed does not allow the source of the defect to be eliminated. Inappropriate methods of monitoring quality may misguide a firm into believing it is delivering acceptable products/services when in fact it is not. A direct result of this is the production of batches of defective products/services.

6. The waste from inventory. Excess production will be transferred to inventory where it runs the risk of becoming damaged or obsolete. Other unnecessary costs included are the costs of excess raw materials and component parts not required to produce the final product. The cost reduction associated with materials is estimated to be 30–50 per cent of total operating costs. These cost reductions include the following:
   a. The elimination of holding inventory. The cost savings from this are threefold: reduction of storage facilities, reduced dangers of obsolescence and of potential theft and damage to the inventory.
   b. The elimination of bulk breaking. This involves the breaking down of large shipments into smaller lots that can be readily used for production purposes.
   c. Reduction in the number of suppliers. JIT requires the use of only a few suppliers. The success of this depends upon the development of a trusting relationship between the customer and the supplier. It also requires supplier dependability with respect to the stock arriving when it is needed in order to fulfill consumer demand.
   d. The development of long-term contracts. This helps to ensure that supplies needed for production will be received. It also removes the risk of a company not being able to negotiate a contract with a supplier in terms favourable to both parties. This helps to create a win-win situation for both the supplier and the manufacturer.
   e. Reduction in receiving inspection. This can also be achieved as the supplier contract establishes and enforces the quality of supplies to be received.
7. The waste of transportation. These are the wastes associated with the movement of materials from inventory to different work stations. This arises from inefficient plant layout.

On the other hand, recent studies have identified modes of waste reduction including 'obvious wastes' such as unneeded processes, excessive set-up times, unreliable machines, re-work and the 'obvious wastes' associated with variability (deTreville and Antonakis 2006). Hopp and Spearman (2004) argued that variability in process times, delivery times, yield rates, staff levels, demand rates and so forth may lead to buffering costs. The implementation of JIT can help eliminate obvious wastes, reduce variability and exchange expensive buffers, for example, inventory, for less expensive ones, for example, capacity. In doing so, a firm can expect to achieve performance improvements in the areas of cost efficiency, conformance quality and delivery speed and reliability. These improvements can result from greater resource productivity and utilization, lower overhead, lower inventories and faster cycle and throughput times.

In addition to the 'value-added' and 'waste reduction' rationale for implementing JIT, there are other reasons behind JIT implementation in firms. Motivated by the operational and strategic potential of JIT for enhancing firms' performance, Yasin et al. (2004) reported the result of a stream of research aimed at gaining practical insights into the different facets of effective JIT manufacturing, service and public sector firms. Using structured and unstructured interviews with some of these sampled firms, they identified similarities and differences among these firms and their respective sectors concerning their JIT experiences. Findings from their study are briefly summarized below.

Reasons behind implementing JIT:

- increase the efficiency of operations;

- improve quality;
- increase customer satisfaction;
- improve management-workers relations; and
- gain a competitive strategic advantage.

Benefits of JIT:

- elimination of some material handlers resulting in labour cost savings as a result of JIT;
- reduction in set-up (ordering) costs;
- reduction in WIP inventory;
- reduction in lead time;
- improvement in the quality level of incoming materials;
- less paperwork;
- significant reduction in rejects of outgoing final products/services; and
- reduction in the number of grievances filed by workers.

Prerequisites to JIT Implementation

Prerequisites to implementing JIT encompass all the actions and preparation that are required of the firm prior to embarking upon JIT implementation. In their study, Yasin et al. (2004) reported on the actual modifications being enacted by manufacturing, service and public firms as they prepare for JIT implementation. The modifications initiated prior to JIT implementation include:

- reduction of the number of vendors;
- changing facility layout and combining operations;
- training employees to improve job skills and problem-solving ability;
- changing job classification;
- grouping machines (equipment) in cells;
- purchasing equipment with short set-up times (preparation time);
- changing inventory and order policies; and
- initiation of quality circles and quality control programmes.

Yasin et al. (2004) also identified a number of factors contributing to the failure of JIT implementation, which include:

- lack of cooperation from vendors in direct linkages with vendors;
- lack of resources to invest in direct linkages with vendors;
- unwillingness of workers to perform multi-tasks;
- management's resistance to sharing operational power with employees;
- lack of management confidence in hourly workers' commitment to the organization; and
- lack of accurate forecasting systems.

Summarizing their study results, it is evident that organizational flexibility is essential for firms planning to adopt JIT. The firm may be required to respond to situations that are very different from those it is accustomed to, as JIT may inflict very new and foreign experiences on the firm. The firm's ability to accommodate these experiences will be measured by its capacity to respond quickly to these experiences and demands.

Firms should consider flexibility on four levels: adjustment to changes in volume, modification of the product mix, choice of equipment and people flexibility (Hall 1987). These are explained below:

1. Flexibility to adjust to changes in volume pertains to the firm's willingness to plan carefully and to analyze future capital expenditures. Capital expenditures should be incurred when such purchase will assist in meeting the purposes of operations and complement the overall manufacturing processes. Flexibility of this nature also implies that an organization should strive to maintain low levels of overhead, process costs and equipment in order to achieve a low break-even point.
2. Flexibility to modify the product mix will require a firm to employ multi-skilled workers, low inventory levels with a wide variety of parts and reduced set-up times for operations.
3. Flexibility in choice of equipment for operations will be a consideration of an organization when it is faced with specific tasks. The first approach to meeting the demand of specific tasks is to adapt the existing general purpose equipment to those tasks. In the event that this approach proves to be inappropriate, equipment designed to perform the specific tasks should then be purchased or built, but at the lowest cost possible to the firm.
4. Development of employees to acquire multiple skills, or of specialists willing and able to accommodate the needs of production, should be focused on as a means of creating an organization with greater flexibility. Employing such people will allow the firm to meet variations in demand and ensure that the production/service operations of the firm can continue to run in a smooth and steady manner. A firm that fails to cultivate flexible employees may be characterized by such occurrences as stoppages in production/service operations. These stoppages can result from employees who are hesitant to perform a necessary task because it is not directly related to their job function.
5. Development of supplier capability to meet the JIT implementation requirement is necessary because suppliers are considered partners with JIT firms. Since the supplier is a vertical extension of a JIT firm's operations system, it is usually necessary for JIT firms to reduce their supplier bases. Suppliers that do not have sufficient capability to meet all of the JIT delivery and quality expectations may need to be eliminated. On the other hand, development and training efforts should be provided to the remaining suppliers to enhance their capability to meet the JIT requirements.

Summary

In this chapter we learnt that JIT is a Japanese management philosophy that has been widely adopted by manufacturing firms since the 1970s. The belief that JIT is also applicable to the service sector has encouraged its deployment among service practitioners but the deployment had not been popular until the mid-1990s. JIT is a management philosophy that emphasizes waste reduction through continuous process improvement and it helps firms to achieve the dual advantages of cost reduction and quality improvement simultaneously. The success of JIT logistics depends on three important tightly integrated elements, namely, people involvement, logistics network, and systems. In order for firms to leverage the strengths of JIT, they must first understand its objectives, and identify and prioritize their goals that are unique to their organizations. On the other hand, they will also need to be aware of the shortcomings of JIT to prevent their operations from unexpected disruptions. If carried out successfully, JIT logistics allows firms to remain competitive by integrating and optimizing activities in their value chains and responding better to their customers' needs.

CHAPTER 3
Basics of Logistics

History and Development of Logistics

The term 'logistics' has long been associated with the military. It was regarded as all the activities involved in wartime deployment and support of a nation's armies. The earliest traces of logistics can be found as early as 500 BC. Alexander the Great was probably the first military leader who chose to carry supplied for the army along instead of living off the land. He used trains to move soldiers and supplies around and built supply depots containing food and fodder across the empire (Gourdin 2006). The value of logistics is highly appreciated in the modern world. Logistics management was vital to the victory of the US army in the Persian Gulf War in 1990 as they were able to airlift over half a million of people and over half a million tonnes of materials and supplies and transfer 2.3 million tonnes of equipment by sea in a few months (Lambert et al. 1998; Christopher 2005).

Johnson et al. (1999) summarized eight driving forces for the development of logistics since the 1950s:

1. Rising transportation costs – shortage of fuel and soaring fuel prices in the 1970s drew the attention of management to deal with the rapidly changing environment.
2. Untapped efficiencies in distributions – little could be done to reduce the cost in the already highly efficient production process but cost reduction in distribution was yet to be explored.
3. Changing inventory distribution patterns (from placing half of the inventories at retailers to keeping 90 per cent of the stock by manufacturers and distributors) – distribution strategies would have to be redefined.
4. Proliferation of product lines – instead of focusing on a number of products, product varieties grew substantially. More sophisticated inventory management techniques were required.
5. Advances in computer and communication technologies – vast amounts of detailed information allowed firms to evaluate and analyze their logistical performance more easily.
6. Ability to analyze suppliers' services with the help of computer technology.
7. Environmental concerns for product packaging.
8. Sophisticated logistics services developed by retail giants – innovated services revitalized the traditional distribution concepts and forced firms to redesign their purchasing and distribution strategies.

The development of logistics has close ties with the history of American business and is largely driven by the demand and supply forces in the market place. Little attention was paid to logistics activities before the 1950s when the demand of goods remained high and exceeded supply. The recessions of the 1950s, which led to a sudden plunge in demand, alerted managers to the need for cost-effective distribution networks. Since then, the notion of logistics started to conceptualize. In the mid-1960s, faster and more reliable distributions at lowered costs encouraged firms to start looking at the possibilities of including order processing, warehousing, transportation and inventory control in logistics. Materials management was also introduced in an attempt to integrate purchasing and manufacturing functions. Resource uncertainties during the recessions of the 1970s forced firms to switch their priority from meeting demand to maintaining both the quantity and quality of supply. Materials management became the operational focus. Development of mainframe computers assisted firms in developing models to evaluate logistical alternatives. From the 1980s onwards, firms once again regarded logistics as a tool to boost productivity and efficiency as they looked forward to take advantage of transportation deregulations and the development of both the microprocessor and communication technologies (Bowersox et al. 1986). Since 1968, many of the government-owned transportation carriers such as rail and airlines were sold to private businesses and thereby shippers have been provided with more choices of services at competitive prices. Nowadays, logistics is viewed as a management approach for organizations to develop sustainable competitive advantage (Gourdin 2006). It is an integral part of a firm's strategy to deliver their products and services to satisfy its customers. Figure 3.1 depicts the development of business logistics from the past to the present.

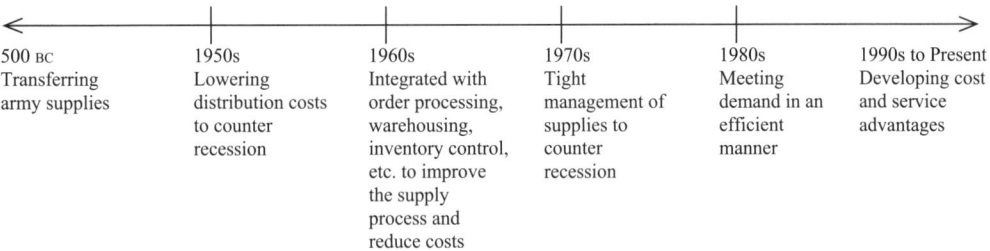

Figure 3.1   The development of logistics

Similar to the JIT management approach, there is no universally accepted definition of logistics. Ballou (1987) considered logistics as: 'All the move-store activities that facilitate product flow from one point of raw material acquisition to the point of final consumption, as well as information flows that set the production in motion for the purpose of providing adequate levels of customer service at a reasonable cost.'

He further defined the mission of logistics as: 'Getting the right goods or services to the right place, at the right time, and in the desired condition, while making greatest contribution to the firm.' (Ballou 2004).

In 1991, the Council of Logistics Management (now the Council of Supply Chain Management Professionals) summarized the definition of 'logistics' as: 'Logistics is the process of planning, implementing and controlling the efficient, cost-effective flow and storage of raw materials, in-process inventory, finished goods and related information from point of origin to point of consumption for the purpose of conforming to customer requirements.'

Though the notion of 'logistics' is being interpreted differently by scholars and practitioners, these definitions do share commonalities. They all focus on the importance of managing raw materials/products throughout the supply chain in an effective (conforming to customer requirements) and efficient (operating at lowest costs possible) manner. Every business firm needs logistical activities to a certain degree. It is an economic fact of life that resources, and the consumers of these resources, are widely scattered geographically. This implies that provision is needed to distribute the resources in ways that ensure consumers can have the products/services they want in the right quantities and at the right place at the right time and in the right condition at the right price with the right information (Coyle et al. 2003).

Improving customer service, making operations faster, more responsive and dramatically reducing costs are the challenges faced by many firms today. Some proactive ones have started to search for ways to improve their ability to reduce waste, improve services and compete globally. Coupled with the philosophical emphases of the JIT management approach, there is a higher chance for logistics management to enable firms to attain both cost and service advantages.

Scope of Logistics

The customary association of logistics with only manufacturing applications is considered 'narrow and inappropriate' by Ballou (2004). Indeed, the logistics concept also applies to the service sector. Government bodies, hospitals, banks, retail shops and schools are typical examples of service sector entities that are involved in logistical activities. Actually, many service firms are offering tangible products. Examples of such firms include fast food giant McDonald's, newspaper publisher Dow Jones & Co., Inc. and retailer Sears. These service operators carry out the same logistics activities as manufacturers. For those that do not offer physical products, logistics activities, though not obvious, exist in their workflow and decision-making processes. For instance, hospitals need to reduce the costs of acquiring medical supplies, express couriers need to design optimal pickup and delivery routes, and banks need to locate and retain cash inventories for their ATMs. Because of the increasing concern for disposal, recycling and reuse of products in the modern business environment, logistics also encompasses activities such as removing packaging once a product is delivered, and removing old equipment.

The Logistics System

The logistics concept is based on a total system view of materials and product flow activity, from the source of supply through to the final point of consumption. A logistics system encompasses different organizational areas and consists of a number of activities that support its operations. A logistics system relies on various resources for input (Stock and Lambert 2001). Financial resources are needed to purchase equipment and machineries and hire people. Human resources are required to plan, manage and carry out the logistics activities. Infrastructure resources including organizational structure, plants and facilities and information systems are also necessary inputs. As logistics encompasses activities across different functional units in a firm, the organizational structure must then be such that it alleviates the coordination efforts among units. Information resources are necessary to facilitate communication among the different parties involved in the logistics process and to enable access to data for decision making.

A logistics system starts with the provision of raw materials, in-process inventory and finished goods by suppliers. The management actions, that is, planning, implementation and control, provide a managerial framework for firms to perform the required logistics activities to attain such business goals as the creation of time and place utility and the reduction of cost. According to Ballou (2004), these logistics activities are termed the 'logistics mix', which can be classified, according to their importance to logistics management, into primary and supporting activities. There are four activities considered to have primary importance in achieving the cost and service objectives of logistics. These key activities include:

1. customer service;
2. transportation;
3. inventory management;
4. order processing.

These four activities are considered key elements in logistics management as they contribute most of the logistics costs in firms and are essential to the effective coordination and completion of the logistics tasks. In support of these primary activities, there are a number of additional logistics activities. These supporting activities include:

1. warehousing;
2. purchasing;
3. materials handling;
4. packaging;
5. production scheduling;
6. information maintenance.

Although these supporting activities may be as important as the key activities in any particular situation to help firms to attain cost reduction and service improvement, they are complementary to the key activities in contributing to the logistics mission of a firm. Figure 3.2 shows the relationships among logistics mix activities in a firm.

Unlike the key activities, not all of them will take place in every logistics process in a supply chain, and one or more of them may not be a part of the logistics activity mix of every firm. For instance, bulk cargo such as iron ore and timber does not require packaging for protection against bad weather and theft, even though inventory management and materials handling are necessary for these items.

It is important to understand that a major goal of a firm is to maximize long-term profitability. One way to achieve this is to adopt a systems view to manage the logistics activities as an integrated system. A logistics system requires that logistics activities do not operate independently, but as part of an integrated system. A firm should also integrate their flows of products/services in their management of logistics from the system perspective. Heskett et al. (1973) made the following observations concerning logistics systems:

- a logistics system possesses multiple interrelated parts;
- the performance of one part affects and is affected by that of others; consequently, to analyze any subcomponent in isolation constitutes a serious methodological error;

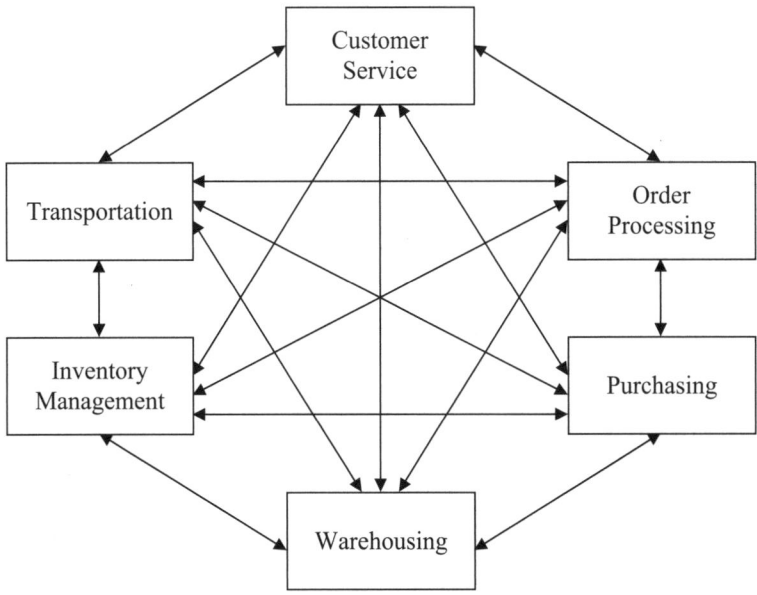

Figure 3.2 The logistics mix

- the alternation of certain subcomponents generates more change in system behaviour than others;
- overall system performance is dependent upon the 'balance' achieved among the subcomponents;
- the weakest member often dictates the upper bound of system performance;
- optimum system performance is often not dependent upon the optimal performance of each subcomponent, but requires balance and coordination among them.

This systems approach requires firms to examine the trade-off alternatives, thereby reducing the overall total cost of logistics activities within a system. For instance, carrying more inventory improves customer service, but at the cost of lower inventory turns. Acquiring inventory in greater quantities per order and increasing the length of order cycle might reduce handling, transportation and purchase costs, but could also create higher inventory carrying costs and/or negatively affect customer service. In firms that do not undertake a system approach, the management of logistics could become a fragmented and uncoordinated set of activities spread throughout various organizational functions, with each individual function having its own budget, set of priorities and performance measures. A logistics system calls for holistic management and attention to the cost aspect in managing logistics functions. Firms should strive to reduce the total cost of all the activities instead of isolated activities as there are always trade-offs in managing the various operations. For instance, if one would like to reduce the inventory carrying costs by maintaining lower inventory levels, higher transportation costs will be incurred as more frequent deliveries are required.

As Stock and Lambert have observed (2001), many firms such as Herman Miller, 3M, Quaker Oats and Whirlpool Corporation found that total logistics costs can be reduced by integrating such logistics-related activities as customer service, transportation, warehousing, inventory investment, order processing and information systems, and production planning and purchasing. Without the system approach to manage logistics, operations inefficiencies such as inventory build-up can result. Because of the lack of a coherent objective for logistics, these efficiencies can easily take place in the following critical business interfaces:

- supplier-purchasing – purchasing management is often rewarded for achieving low per-unit costs for raw materials and supplies.
- purchasing-production – production management is usually compensated for achieving the lowest possible per-unit production costs.
- production-marketing – salespeople like to have market presence by positioning large inventories of product in the field, as close to the customer as possible. This allows salespeople to offer the shortest possible order cycle time and to minimize the difficulties associated with forecasting customer needs.
- marketing-distribution – in some companies, transportation is the only logistics cost that is closely monitored. Transportation managers have more incentive to ship products by truckload or by railcar in order to obtain the lowest possible freight rates. Generally, these large shipments of products require increased inventories at both origin and destinations (for example, manufacturer, wholesaler and retailer).
- distribution-intermediary-customer – both customers and intermediaries (for example, wholesaler and/or retailer) may attempt to reduce their inventories by purchasing more frequently, thereby forcing their inventories and their associated carrying costs back towards manufacturers. This is particularly true when intermediaries are concerned about cash flow.

Inbound and Outbound Logistics

The activities that make up the supply chain process vary from firm to firm, depending on a firm's particular organizational structure, the management perception of the supply chain scope of the firm and the importance of individual activities to its operations. Ballou (2004) viewed that there are two aspects of logistics activities in a firm's immediate supply chain, that is, inbound logistics (materials management) and outbound logistics (physical distribution). Inbound logistics is concerned with products (for example, raw materials, spare parts, assembles) moving into a firm rather than away from it. The inbound logistics activities are those activities involving receiving, storing and disseminating inputs to the operations areas. The thrust of inbound logistics is to satisfy the operations needs of the manufacturing and service operations line. Outbound logistics is the flip side of inbound logistics, dealing with the movement, storage and processing of orders for a firm's outputs. The main activities are primarily concerned with finished and semi-finished products, that is, products that a firm offers for sale, and for which the firm plans no further processing.

The objective of inbound logistics is to meet the needs of a firm for inbound items in an orderly, efficient and low-cost manner. On the other hand, the objective of outbound logistics is, subject to a specified level of customer service, to minimize the cost involved in physically moving and storing the items from their point of production to the point where they are delivered. Except for purchasing, the logistics activities that make up inbound and outbound logistics are very similar. The ultimate goal of logistics is to integrate and coordinate all the organizational actors in materials management and physical distribution in such a way that the end markets are served in the most cost-effective manner.

The Core Logistics Elements

Customer service, order processing, inventory management and transportation are considered the four core elements for effective logistics management. This is because they either contribute most to the total cost of logistics or they are essential to the effective coordination and completion of logistics tasks.

- Customer service refers to 'the quality with which the flow of goods and services is managed' (Ballou 2004). It is about getting the right product to the right customer at the right place, in the right condition and at the right time, at the lowest cost possible. Customer service reflects the quality with which the flow of goods and services is managed. It comprises all the order-related performance values offered to customers by the firm. Customer service represents the output of the logistics systems and plays an important role in creating, developing and maintaining customer loyalty and customer satisfaction because it is where the customers directly experience the products and services provided by the firm.
- Order processing involves all the activities concerned with collecting, checking, entering and transmitting sales-order information. It is the means by which suppliers and firms exchange order information. The information collected will provide useful data for market analysis, financial estimation, logistics operations and production. Order processing is considered a primary logistics activity because there is a critical time element in getting goods and services to customers, and it triggers product movement and service delivery. The challenge to the firm is to compress the order cycle time, that is, the time a customer transmits an order to the time the customer receives that order. In today's highly competitive business environment, this is a critical dimension of product/service differentiation.
- Inventory management is about keeping stock levels as low as possible while providing the desired level of stock available to serve customers' demand. Effective and efficient inventory control can lead to significant reduction in logistics costs. Inventory is vital to the success of a firm as it is directly related to all the movement and information flows and thereby has an impact on customers, suppliers and the firm's major functional departments. The challenge of inventory management lies in the firm's ability to determine when various items should be ordered, how much to order and how often to meet customer demand while minimizing the respective costs.
- Transportation refers to the various methods for moving products between different parties in the supply chain. While inventory management adds 'time' value to a product, transportation adds 'place' value. Managing transportation is concerned with selecting and utilizing the appropriate modes, routings and so on. Transportation is important, as no modern firm can operate without providing for the movement of its raw materials and/or finished products in some ways. On the other hand, improved transportation management can lead to increased sales, increased market share, and ultimately to increased profit contribution and growth (Ballou 2004).

Supply Chain Management

The emergence of the global economy and intensified competition have led many firms to recognize the importance of managing their supply chains for fast product introduction and service innovations to the markets. Continuous striving for cost and service advantages has led many firms to structurally combine into global supply chains. Traditionally, firms prepared themselves for competition at the organizational level. Increasingly, they recognize the importance of developing the knowledge, skills and strategies necessary to compete at the supply chain level. Many firms have broadened the scope of their logistics operations to embrace SCM to further improve their cost and service performance. The management of multiple relationships across the supply chain is referred to as SCM (Lambert 2001). SCM is concerned with managing the upstream and downstream relationships with suppliers and customers to deliver customer value at the least cost to the chain as a whole (Christopher 2005). Successful SCM requires coordination and integration of both internal and external business processes. It calls for integration of buyers' and suppliers' decision-making processes with the goal of improving materials flow throughout the supply chain (Wang and Cheng 2007). Effective management of the supply chain is viewed as the driver for reductions in lead times and materials costs, and improvements in product quality and responsiveness (Lai et al. 2002). Generally, there are eight key processes that make up the core of SCM (Croxton et al. 2001). These processes include:

- customer relationship management;
- customer service management;
- demand management;
- order fulfillment;
- manufacturing flow management;
- supplier relationship management;
- product relationship management;
- returns management.

This management approach is not confined to a chain of business with one-to-one, business-to-business relationships. Rather, it covers a network of multiple businesses and relationships, offering the opportunity to capture the synergy of inter-organizational management for greater cost reduction and service improvement. The work processes of SCM bring together suppliers, manufacturers, transporters, retailers and customers for a dynamic but constant flow of information, products and funds in the supply chain that adds value to customers and other stakeholders (Lambert et al. 1998). One of the first articles to use the term SCM was by Houlihan (1985), according to whose descriptions, SCM:

- views the supply chain as a single entity rather than fragmented functions responsible for such areas as purchasing, manufacturing, distribution and sales;
- requires and depends on strategic decision making: supply is a shared objective of every function in the chain and is of strategic importance because of its impact on costs and market share;

- provides a different perspective on inventories: inventories are used as a balancing mechanism of last, not first, resort;
- requires a new approach to systems integration.

There are distinct differences between business logistics and SCM. The former refers to activities that occur within the boundaries of a single firm, while the latter refers to networks of firms that coordinate their actions to deliver product or service offerings to the market. Logistics includes all the activities involving the moving of products and information to, from and between members of a supply chain comprising suppliers, manufacturers, carriers, distributors, retailers and customers. Partner firms in a supply chain join forces in managing their logistics activities, which are aimed at bringing goods, services and information efficiently and effectively to the ultimate customers. Major manufacturers and retailers, such as Hewlett Packard and 7-Eleven, have successfully applied sound SCM principles and best practices to their partner firms to achieve improvements in performance throughout their supply chains.

A single firm is generally not able to control its entire product flow from sources of raw materials to points of final consumption. These activities need to be supported by other intermediaries such as suppliers and distributors in the supply chain instead. SCM would help firms align the objectives of its intermediaries in the chain, ensuring that all parties involved in the logistics processes are working towards the common goal of creating the maximum value to customers at the lowest cost possible. The processes across the supply chain and the development of supply chain relationships are characteristics of SCM. The focus is on strengthening the overall supply chain rather than on suboptimization in individual firms. Therefore, SCM views the supply chain and the firms in it as a single entity, bringing a systems approach to understanding and managing different logistics activities needed to coordinate the flow of products and services to best serve the ultimate customers (Simchi-Levi et al. 2008). This systems approach provides a useful framework for logistics management in which to best respond to business requirements that otherwise would seem to be in conflict with one another. A link in the supply chain may, for instance, incur extra costs in the form of excess inventories of semi-finished products, in order to bring down the total cost of the supply chain as a whole. This is congruent with the practices in JIT manufacturing, where the manufacturing plant can often benefit from postponing the procurement of components until the time when specific customer orders are placed. It is only when the entire supply chain is considered as a larger picture that ways can be found to effectively balance the requirements of the different links in the chain.

The implementation of SCM demands close integration of the internal functions and logistics activities within a firm and effective linkage with the external operations of member firms in the supply chain (Lee 2000; Kannan and Tan 2005). The goal is to achieve improved customer service at reduced overall costs. SCM uses information to manage inventory throughout the channel from the sources of supply to the end-users. The information on the quantity and timing of parts must be accurate as one proceeds from the retail end of the supply channel upstream to the sources of raw materials. The challenge in managing logistics is the ability of firms to devise a strategy in the face of increasing uncertainties in the environment and SCM increasingly becomes the key to developing such a strategy.

Vokurka and Lummus (2000) suggested that firms that have implemented JIT elements that align closely with the supply chain elements can more easily implement their corresponding supply chain elements. Other elements in SCM, such as information visibility across supply chain nodes, which may not align with a specific JIT element, may be more difficult for the company to implement. Table 3.1 compares the elements of JIT and SCM.

The Supply Chain Operations Reference (SCOR) model, which considers the performance requirements of partner firms in a supply chain, also sheds light on the importance of a close working relationship beyond traditional organizational boundaries, with an emphasis on inter-organizational networking across the supply chain (Hwang, Lin and Lyu 2008). The challenge for firms in embracing SCM is to work beyond organizational boundaries and involve member firms in the supply chain to improve costs and services in the chain as a whole. Looked at from this perspective, business logistics in an individual firm has a narrower scope. To achieve greater cost reduction and service improvement, individual firms need to cultivate better coordination with partner firms to support their logistics activities, and this gives rise to the idea of SCM.

Table 3.1 Elements of JIT and supply chain management

Elements of JIT	Elements of SCM
Cross-functional employee training	Cross-functional thinking
Connected manufacturing cells	Connected supply chain nodes
Small batch production	Replenishment for production
Reduce inventory investment	Develop target inventory at each node
Quick equipment changeover	Short cycle times
Total productive maintenance	Reliable systems at each node
Close supplier partnerships	Integrate with suppliers
Close customer partnerships	Fulfil customer requirements
Scheduled synchronized to demand	Synchronized supply and demand
Quality at the source	Reliable nodes with quality at each node
Control production via Kanban	Improve visibility via demand pull
Pull production	Produce to meet demand
Produce some of each product each day	Produce 90 per cent of product every day
Continuous improvement	Review and improve based on performance
Visual control	Information sharing throughout the supply chain
Uniform production levels	Same schedules throughout the supply chain
Pull product from suppliers	Integrate logistics in the supply chain

Recently, Gunasekaran et al. (2008) advocated the idea of Responsive Supply Chain (RSC). As businesses in the twenty-first century have to overcome the challenges of satisfying the demand of customers for products of high quality at low prices, it is highly desirable for firms to respond to customers' unique and rapidly changing needs. Many of them are seriously exploring the potential of developing an agile supply chain to get their products to the market faster at a minimum cost. Considering the significance of both Agile Manufacturing (AM) and SCM for firms to improve their performance, they analyzed both AM and SCM and proposed a framework for RSC on the basis of the characteristics and objectives of AM and SCM. The RSC framework, summarized in Table 3.2, can be employed as a competitive strategy in a network economy in which customized products/services are produced by virtual organizations and exchanged using e-commerce.

The Goals of Logistics

The goals of logistics are concerned with the achievement of the 7Rs. Logistics adds value to products by creating utility. Utility is 'the value or usefulness that an item or service has in fulfilling a want or need'. There are four kinds of utility: form, possession, time and place. Logistics contributes to the creation of time and place utilities. Time utility is being able to provide the product when it is needed whereas place utility is being able to deliver the product to where it is needed (Stock et al. 2001). The goal of logistics is

Table 3.2 Comparison of lean supply chain, agility and RSC

Objectives and major determinants	Lean supply chain	Agility	RSC
Objectives/goals	Reduced costs, moderate speed and flexibility	Increased speed and flexibility. Cost is not a major criteria	Reduced costs, increased speed and flexibility
Strategic planning	Few suppliers, outsourcing IT	Core competencies, global outsourcing, virtual enterprises	Supply management, strategic alliances, virtual enterprises, global outsourcing and IT
Organizational structure	Supplier development	Virtual enterprise, partnership formation based on core competencies	Virtual enterprises, supply chain integration and IT
Knowledge and information technology	Supply chain integration, knowledge workers	Agile and knowledgeable workforce, enterprise resource planning systems	Training and education to operate in a global environment, enterprise resource planning systems

to drive down the total costs of all the activities, or to optimize the total costs incurred, without affecting qualities. The survival and growth of a firm depends on its effectiveness and efficiency in satisfying the demands of its customers and other stakeholders. The combined effect of effective (for example, good services) and efficient management (for example, low operations costs) can be reflected in a firm's balance sheet and its profit and loss account. The double-edged impact of logistics in bringing about both increased profitability and reduced costs on a lower asset base can provide a substantial leverage on the Return on Investment (ROI) of the firm (Christopher 2005).

Consider the two elements that make up ROI, that is, profit and investment. To improve the first dimension of profit, generating high revenue and managing low-cost operations are important. Firms can attain high profit by improving sales revenue through better customer service on one hand and attaining low costs by better control of operations expenses, for example, in inventories, warehousing, transportation and so on, on the other. The second dimension of ROI is capital employed, which represents the investment, which is also sometimes referred to as the 'asset base'. The management of the asset base is as important as the management of profit. Managers are often so engrossed in trying to increase profits through increasing market share that they tend to overlook the management of assets. For instance, poor maintenance of warehouses and materials handling equipment can result in delays, which affect the order cycle, and hence, customer service. Usually, such delays are not reflected in the financial statements, and by the time the effects are recognized, it is perhaps a little too late.

Logistics and Competitiveness

In order to reap the full benefits of business logistics, one should examine how it will enhance a firm's competitiveness. Porter (1985) categorized the activities in the value chain into primary (inbound logistics, operations, outbound logistics, marketing and sales, and services activities) and support (procurement, technology development, human resources management and firm infrastructure) activities. Porter's value chain can be used to illustrate how business logistics can be applied across system activities and subsequently provide the mechanisms for working through traditional organizational barriers, and allow for the integration of activities and create a CI effort throughout the value creation process.

Porter (1985) suggested that for a firm to establish sustainable competitive advantage, it should concentrate its efforts on optimizing or coordinating all the activities in its value chain. The value chain is a series of interdependent activities and this collection of activities are performed by a firm to design, produce, market, deliver and support its products; in other words, a firm is virtually a collection of interdependent activities that are carried out to produce its products. These activities can be broadly grouped into two categories, namely primary and support activities. Primary activities include inbound logistics, operations, outbound logistics, marketing and sales and service; whereas support activities encompass a firm's infrastructure, human resource management, technology development and procurement. All these activities are to various extents related to business logistics.

Reducing costs or improving performance in any of the activities alone will not necessarily guarantee a sustainable competitive edge as either a price cut or enhanced service alone can easily be imitated by competitors. The goal of logistics, therefore, is to optimize the performance of all the activities in this value chain, that is, delivering quality products at low costs, to establish a sustainable competitive advantage. To create a sustainable competitive advantage, firms should concentrate their efforts on optimizing or coordinating all the activities. Reducing costs or improving performance of any of the activities alone will not necessarily guarantee a competitive edge. Logistics represents three of the primary pillars, namely inbound logistics, operations and outbound logistics, of the value chain, which in turn implies that a firm's competitiveness is connected with its ability to manage its logistics processes successfully. The logistics mix activities are becoming a form of proprietary asset of the firm, as well as its source of engendering a competitive edge, since it is difficult for rival firms to imitate the functional structure, supplier and carrier relationship established, and so on, in the logistics processes.

Summary

In this chapter we learnt that logistics has its roots in the military sector. It started to draw considerable attention from the business sector in the 1950s. Today, logistics refers to all the move-store activities from the point of raw materials acquisition to the point of final consumptions. Logistics creates value for firms by providing time and place utilities for their products/services. Its core elements include customer service, order processing, inventory management and transportation. Each of these elements is critical to the successful operation of a firm's logistics process and they are supported by other logistics activities such as information management, materials management and so on. Realizing that either price cuts or superior products alone cannot guarantee a sustainable competitive edge, and in the face of the ever-increasing uncertainties in the market place, SCM has become the key to cope with these challenges. SCM requires intermediaries or partners to engage in cooperative relationships and to work in harmony to deliver quality products at low costs.

JIT and logistics represent alternate approaches to improving the effectiveness and efficiency of a firm's operations function. While there are differences in their motivations and objectives, both JIT and logistics seek improvements in quality, the former by way of improvements in production processes, the latter in the logistics processes. Successful JIT implementation depends on the coordination of production schedules with supplier deliveries, and on high levels of service from suppliers, both in terms of product quality and delivery reliability. This requires the development of close relations with suppliers and the integration of a firm's production plants with those of its suppliers. It can be surmised that while the two approaches have certain defining characteristics, they represent elements of an integrated operations strategy. The issues concerning the integration of these two management approaches and how JIT logistics can benefit firms by improving their business performance are discussed in the next chapter.

CHAPTER 4
JIT Logistics

Many firms have found JIT logistics, with its focus on business logistics, difficult to practise. From an operations perspective, the scope of business logistics is very broad. Wastes in business logistics can take place in many different instances. Examples include unnecessary expedite shipments due to a lack in the visibility of shipments and requirement schedules, excessive production due to inaccurate forecast of market demand and lost sales due to misunderstanding or slow response to customer requirements. For a better understanding of JIT logistics, it is helpful to break down logistics into smaller logistics mix elements. By doing so, we can create a theoretical foundation to discuss the principles and perspectives of JIT logistics.

In recent years many innovative management concepts and terms have emerged in the business literature that builds upon the basic ideas of JIT logistics. These concepts focus more on the two-sided relationships between customers and suppliers. SCM, Customer Relationship Management (CRM), Vendor Managed Inventory (VMI), Efficient Customer Response (ECR), and Cooperative Planning, Forecasting and Replenishment (CPFR) are the most widely known among these new management concepts.

In the following sections we will discuss how the JIT concepts can be applied to each of the four primary logistics mix elements, namely customer service, order processing, inventory management and transportation management. We will also discuss the emerging management concepts in the context of JIT logistics.

Customer Service

Customer service represents a critical element in business logistics. It is the final output, which can be a physical product (for example, a soft drink), a service (for example, postal service) or a combination of both (for example, a fast food restaurant, medical services), of the entire logistics process, and has a direct impact on logistical performance and customer satisfaction. Customer service in business logistics is concerned with the quality with which the flow of goods and services is managed and with the mission of business logistics in essence. Attention to customer service should not only be limited to service levels, for example, availability of goods to customers, but also the financial costs incurred from providing quality customer services because minimizing costs is key to survival and prosperity in a competitive marketplace.

JIT customer service requires fundamental changes in the way in which business is conducted both internally and across organizational boundaries. It is challenging to fulfil the JIT logistics' objective of waste reduction by delivering the right product in the right quantities at the right time. The 'right product' means focusing not only on delivering a product that conforms to specifications, but also on the associated value-added services and the processes that produce them. This translates into evaluation and selection of JIT partners, less reliance on inspection of inventories (at the exchange points in the supply chain) and joint involvement of supply chain members in new product or material designs. In the 'right quantities at the right times' means that customers get exactly what they want or need at the time they want or need it. Precise quantity and timing can be derived from dependable estimates of demand from buyers, who involve suppliers to a large degree in their inventory replenishment decisions and who electronically share information about their product flows (for example, through EDI). Precise quantities and timing have many implications, including on-time delivery, smaller lot sizes and more frequent shipments of variable sizes. The result is that smaller inventories are held at every point in the supply chain (and, internally, at every stage of the production/warehousing cycle). Demand for smaller, more frequent shipments also makes geographic proximity of partners a competitive advantage for a supply chain. To understand customer service in logistics more fully, it is useful to explore its meaning and elements, and their impacts on customer satisfaction in more detail.

MEANING OF CUSTOMER SERVICE IN LOGISTICS

The term 'customer service' is omnipresent in our daily lives, such as 'customer services hotline', 'customer services counter' and 'customer services representative'. What is customer service in the logistics context? One way to decode its meaning is by listing the various ways in which customer service has been categorized. The list, in order of popularity, as reported by Ballou (1987), is:

1. the elapsed time between the receipt of an order at a supplier's warehouse and the shipment of the order from the warehouse;
2. the minimum size of the order, or limits on the assortment of items in an order, that a supplier will accept from its customers;
3. the percentage of items in a supplier's warehouse that might be found to be out of stock at any given point in time;
4. the proportion of customer orders filled accurately;
5. the percentage of customers served, or volume of orders delivered, within a certain time from the receipt of the order;
6. the percentage of customer orders that can be filled completely on receipt at a supplier's warehouse;
7. the proportion of goods that arrive at a customer's place of a business in saleable condition;
8. the elapsed time between the placement of an order by a customer and the delivery of the ordered goods;
9. the ease and flexibility with which a customer can place an order.

Although there are different perceptions of customer service in logistics, they place a common emphasis on satisfying customer requirements. In recent years, retaining customers has become a challenge for many firms due to rising customer expectations and intense competition. Determining customers' needs and delivering services that meet those needs in a cost-effective way is a key concern for logistics management. As customer service is the output of a logistics system, it is playing an increasingly important role in attaining customer satisfaction. Yet, there are other product/service attributes such as product quality, pricing and promotion that a firm can use to retain customers. These tangible attributes are easy for competitors to copy. What if the competitors manage to approach these attributes and erode these competitive strengths of the firm? It is probable that customer service in logistics can be the differentiator and can help to bring customers back. Generally, it is rather difficult for competitors to exactly duplicate the logistics management practices of a firm, for example, fast delivery and responsive handling of customer complaints. For instance, Amazon.com competes with physical stores on book sales in price and variety. Nevertheless, Amazon.com can survive even with the existence of all its competitors because of its good customer services, for example, product availability, product variety and after-sales services, even though its competitors can get close to meeting, or even beating, its product prices.

One important note is that firms need to adopt different customer service definitions to cater for varied customer needs in different contexts, for example, manufacturing, distribution, shipping and retailing. If a firm wants to achieve a desired level of customer service, it is important that the firm has an agreed definition of customer service with the customers and suppliers in its supply chain. Simply put, the actual customer service level provided must match the service level customers expect. This consideration is important for delivering consistent customer services (between firms and suppliers) at the level desired by customers. For example, Lenovo may intend to perform at a high customer service level, for example, rendering installation services one day after sale for its personal computer products, and its customers may well desire and expect this level of service. But if Lenovo's suppliers and distributors fail to deliver this service, customer dissatisfaction is likely to take place.

CUSTOMER SERVICE ELEMENTS

Customer service in logistics encompasses a number of service elements before, during and after actual sale transactions. There are three generic customer service elements as categorized by Ballou (2004). These categories include pre-transaction elements, transaction elements and post-transaction elements, which are detailed below.

Pre-transaction elements involve cultivating an atmosphere for good customer service. Generally, these activities are related to a firm's policies regarding customer service, and can have a significant impact on customers' perceptions of the firm and their overall satisfaction. Examples include when goods will be delivered after an order is placed, the procedure for handling returns and back orders, and methods of shipments. It is also desirable for firms to develop plans for the time when labour strikes or natural disasters affect normal service, create organizational structures to implement customer service policies and provide technical training and manuals to cultivate better buyer-seller relationships.

Transaction elements are those that directly result in the delivery of a product/service to a customer. Examples include determining the level of availability of the stock, selecting transportation modes and establishing order-processing procedures. These elements can affect the performance in order delivery, accuracy of order filling, condition of goods on receipt and stock availability.

Post-transaction elements refer to the variety of services needed to support the products/services in the field. Examples include handling of defect products, return of packaging, complaint handling and claims processing. Although these activities take place after the sale of products/services, they must be planned for in the pre-transaction and transaction stages.

OBJECTIVES OF CUSTOMER SERVICE FOR JIT LOGISTICS

The desired outcome of JIT logistics is customer service. In order to provide the desired level of customer service, firms must identify customer requirements and then provide the right combination of transportation, storage, packaging and information services to successfully satisfy those requirements. The goal of JIT logistics is to deliver the right product to the right customer at the right place in the right condition at the right time. JIT logistics helps to achieve this goal by creating four forms of value for the customers, namely time, place, form and possession utilities.

Time utility – providing customers with time utility means that customers are able to obtain what they desire at the designated time (for example, fast-moving consumer goods are expected to be available on the shelves of grocery stores all the time, room services at hotel rooms are expected to be delivered as soon as possible). To guarantee that a product or service reaches the customers at the desired point in time, firms need to exercise tight control of the duration in each step in the logistical process. Order cycle timeliness is no doubt one of the key processes that firms need to keep track of as the response time of a firm to a customer's order is mostly made up of the time taken by the order cycle process. As to be discussed later in this chapter, under JIT, firms will seek close cooperation with suppliers in the form of long-term partnership programmes and provide them with future demand forecasts. The accuracy of forecast models relies on the use of information systems that gather and analyze previous sales history, as well as alignment with key customers to generate ideas for future production. Besides initiating closer relationships with suppliers and customers, JIT also suggests firms ensure that raw materials and WIP items arrive at the time when they are needed, finished goods are delivered at once and workers with similar or related tasks are seated within close proximity of each other so that motion time is minimized. Since we are going to discuss in the following sections how JIT improves order cycle management, inventory management and transportation management, which all affect the timeliness of a firm's response to an order, details of JIT practices are not elaborated for the time being.

Place utility – the notion of 'providing the right product at the right time at the right place' explicitly pinpoints the importance of delivering an item to the designated spot, which is, in essence, providing place utility. Allowing customers to obtain a firm's product or service at the spot where they want is termed place utility. The ability of a firm to deliver its products to various locations relies heavily on its transportation network. Similar to managing a supplier network, firms are encouraged to maintain cooperative relationships with their carriers. Advances in technology also allow firms to analyze their delivery networks and develop delivery routes that will serve to transport the items in good shape in a timely and cost-effective fashion. We will examine how JIT improves transportation management later in this chapter.

Form utility – providing form utility means that a firm is converting inputs into finished products for customers' consumption. Under JIT, manufacturing products or providing services is not sufficient; it is about producing the right product, that is, producing a product that conforms to the specifications from a defect-free process. JIT emphasizes continuous process improvements to achieve the goal of zero defects. Everyone in the organization is responsible for quality control. Workers are encouraged to voice out any inefficient or ineffective areas in the production process and are empowered to halt production if any discrepancies are spotted.

Possession utility – possession utility is helping customers to acquire the products that they desire. If firms are able to provide time and place utilities, possession utility is offered at the same time since customers can obtain what they desire at the designated spot and time with ease. The objective of customer service for JIT logistics is simple and straightforward, namely to satisfy customers by delivering the right goods or services in the right quantities at the right time while minimizing total process cost by eliminating waste of all kinds from the supply chain. This objective can be accomplished effectively by taking the following steps: 1) developing relevant policies and procedures for managing the supply chain as a whole; 2) recognizing the service level requirements of the final customer; and 3) determining where to position inventories along the supply chain and the level of inventory stock at each point. The coordination and integration of all the parties involved in the supply chain is critical to delivering customer service from the successful development and implementation of JIT logistics.

WASTES IN CUSTOMER SERVICE

Under JIT logistics, waste in customer services is anything that adds no value to customers in the logistics processes. Below are some examples of possible wastes that can be identified in the customer service area:

- Untrained service personnel – service personnel without proper training cannot guarantee to attain a reasonable service level (for example, insufficient product knowledge) and are likely to turn prospective or current customers away.
- Service personnel do not have authority to alter or adjust the service process – customers will not wait indefinitely for a firm's response to their enquiries or problems. If front-line employees do not have the authority to provide satisfactory solutions on an instantaneous basis, customers are lost forever.
- Unavailable products – if the product or service is not available or the quantity is not sufficient for consumption at the customer's desired point in time, the loss of sale becomes a waste. Loss of sale is costly to firms because they do not simply lose a single sale transaction, customers may switch to their competitors for the same product or service and do not return.
- Service discrepancies – late deliveries, defective or damaged products and delivered items not matching the orders placed are examples of service discrepancies that will affect customers' perception towards a firm's service quality.
- Inadequate or insufficient product information – if customers are not provided with comprehensive product information, their purchase decisions may be affected and loss of sales may result. Customers will be dissatisfied if the information provided does not match the actual product or service specifications.
- Delayed responses in handling order discrepancies – if order discrepancies unfortunately occur, firms should strive to react quickly to rectify the problems, as delayed responses will further disappoint the already annoyed customers.

To reduce wastes, firms should not chase service levels that they cannot afford to support. Simply put, firms should avoid spending on service that does not promise a good business return. There are always ways that service can be reduced that makes little or no impact on the customers. An example to illustrate is Domino Pizza, which no longer guarantees pizza delivery within 30 minutes, but the firm still remains the industry leader in fast delivery service. Firms exist to offer higher service levels to all the customers all the time, where customers are always asking for more and competitors are always offering more. It is important for firms to determine proper service levels and to avoid wastes. Several challenges need to be overcome: 1) clearly understand how service is positioned within the firm's overall business strategy; 2) understand the costs of each component of the service offering; and 3) know each customer's need in enough detail to avoid spending on services that are not valued.

A good point to begin with is determining service-level objectives. Some prompting questions may include: Does the firms' strategic objectives require industry norms for service offerings? Or is service the core element that will set the firm apart from all others in its market? Once the service objectives are determined, the firm can establish appropriate offerings and formulate related policies and strategies. For instance, the emphasis of Wal-Mart is on value for money, not premium service. Lowes puts a lot of resources on sales-floor associates, product specialists and project planners because service is viewed as a key differentiator. Neither Wal-Mart nor Lowes spends unnecessarily on its service objectives.

Typically, there are three causes for firms to create wastes in customer services: unrealistically high goals, imperfect measures to track service and no business case to support specific service levels. Customers want only and exactly what they want. Anything more than what the customers want is waste, and anything less is poor service. To reduce wastes in customer service, it is useful to ask questions about your customers. What kind of customers does your firm serve? Who are the target customers of the products or services of your firm? Which part of the supply chain is your firm in and what are the customer requirements to sell that part of the chain. Chopra and Meindl (2007) identified the following attributes that help to clarify requirement for different types of customers:

- the amount of the product needed in each order;
- the response time that customers are willing to wait;
- the variety of products needed;
- the service level required;
- the price of the product;
- the desired rate of innovation in the product.

The importance of managing customer service based on cost and customer feedback should not be neglected. By segmenting customers according to their needs and truly understanding those needs, service can be matched much more closely to the goal of waste reduction under JIT logistics.

CUSTOMER EXPECTATION AND CUSTOMER SATISFACTION

To determine customer service levels, firms need to decide on the level of performance that their customers expect. A high-quality service is one that performs at a level that matches the level that the customer feels should be provided (Asubonteng et al. 1996). Simply put, service quality is nothing more than a comparison of what customers feel a firm should offer with their perceptions of what is actually offered. *Customer expectation* refers to the desires or wants of customers, that is, what they feel a service provider *should* offer rather than would offer (Parasuraman et al. 1988). To improve quality is to enlarge the differences between customer expectation and perception, that is, to exceed customer expectation (Grapentine 1998).

Customer satisfaction is often regarded as an indicator of the appreciation of the service provided. It is a confirmation or relates to confirmation of customer expectation (Ham et al. 2003). What makes a satisfied customer? Service quality has long been proven to be one of the drivers for customer satisfaction (for example, McDougall and Levesque 2000). Satisfied customers bring in sales, generate revenues and fuel growth.

Different from product quality, which can be measured precisely, for example, defects per certain number of items, service quality is much more of a subjective perception and is difficult to evaluate. In 1985, Parasuraman et al. (PZB) published a conceptual paper identifying five service quality gaps, which are shown in Figure 4.1:

- Gap 1 – difference between consumer expectations and management perceptions of customer expectations;
- Gap 2 – difference between management perceptions of customer expectations and service quality specifications;
- Gap 3 – difference between service quality specifications and the service actually delivered;
- Gap 4 – difference between service delivery and what is communicated about the service to consumers;
- Gap 5 – difference between consumer expectations and perceptions.

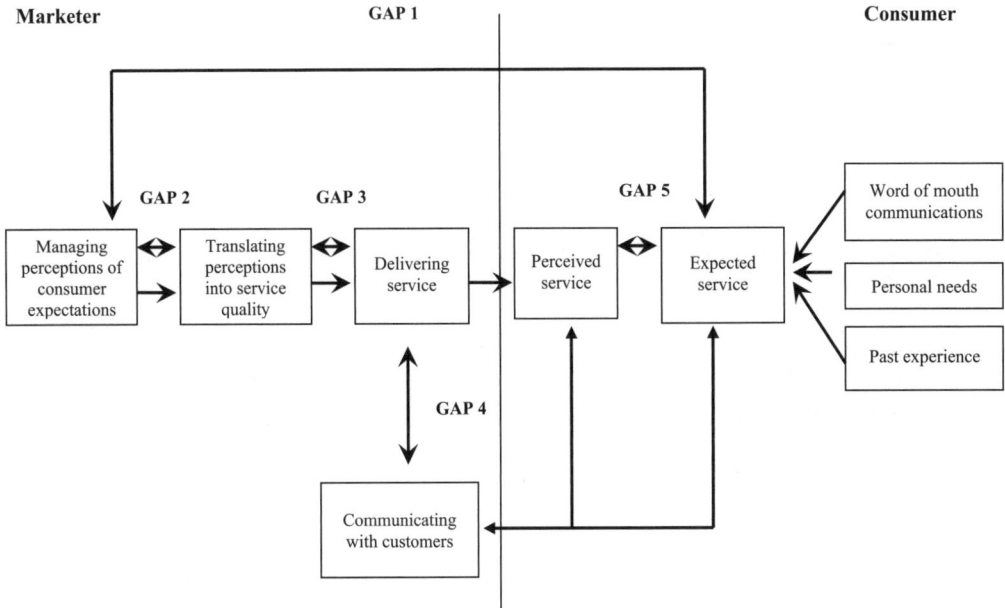

Figure 4.1 Service quality model

Of particular interest to PZB was Gap 5 – the expected service/perceived service gap, which was the focus of their proposed SERVQUAL instrument published in 1988. SERVQUAL is a 22-item measurement scale that helps service firms, and logistics services included, in different sectors to manage customer expectations and perceptions. The scale can be administered to customers and request them to relate the statement in each of the 22 items to their feelings about a company, say, ABC. The respondents will show, in the scale, the extent to which they believe ABC has the feature described by the statement. These 22 measurement items are listed below:

1. ABC has up-to-date equipment.
2. ABC's physical facilities are visually appealing.
3. ABC's employees are well dressed and appear neat.
4. The appearance of the physical facilities of ABC is in keeping with the type of services provided.
5. When ABC promises to do something by a certain time, it does so.
6. When you have problems, ABC is sympathetic and reassuring.
7. ABC is dependable.
8. ABC provides its services at the time it promises to do so.
9. ABC keeps its records accurately.
10. ABC does not tell customers exactly when services will be performed.
11. You do not receive prompt service from ABC's employees.
12. Employees of ABC are not always willing to help customers.
13. Employees of ABC are too busy to respond to customer requests promptly.
14. You can trust employees of ABC.
15. You feel safe in your transactions with ABC's employees.
16. Employees of ABC are polite.
17. Employees get adequate support from ABC to do their jobs well.
18. ABC does not give individual attention.
19. Employees of ABC do not give you personal attention.
20. Employees of ABC do not know what your needs are.
21. ABC does not have your best interests at heart.
22. ABC does not have operating hours convenient to all their customers.

The 22 items in SERQUVAL can be categorized into five dimensions that PZB claimed customers will rely on to evaluate service quality. The dimensions are:

1. Reliability – ability to perform the promised service dependably and accurately;
2. Tangibles – physical facilities, equipment and appearance of personnel;
3. Responsiveness – willingness to help customers and provide prompt service;
4. Assurance – knowledge and courtesy of employees and their ability to inspire trust and confidence;
5. Empathy – caring, individualized attention the firm provides its customers.

These measurement items can be used to measure customer expectation and perception on the five dimensions. Parasuraman et al. (1997) argued that collecting customer opinions continuously via a service quality information system allows firms to identify the deficiencies in their services, thereby offering them helpful clues on improving their service quality.

Logistics Service Quality

The 22-item SERVQUAL measurement scale is designed to cater for general service settings, which may not be readily applicable to logistics-oriented sectors. Researchers and practitioners find it necessary to develop a more specific instrument to manage customer expectations in the logistical context. Further to SERVQUAL, Mentzer et al. (2001) presented nine important components of logistics service quality. These components are:

- *Personal contact quality* – personal contact quality refers to the customer orientation of the contact people. Do customers perceive the customer service personnel knowledgeable? Do the service personnel show empathy with customers' situations? Can they help solve customers' problems?
- *Order release quantities* – order release quantities are associated with product availabilities. Customers should be satisfied if they are able to obtain the quantities that they desire.
- *Information quality* – information quality is customers' perceptions of the information provided by the supplier on the products that can be chosen. Product information should always be available and of adequate quality to help customers make their purchase decisions.
- *Ordering procedures* – ordering procedures relate to the efficiency and effectiveness of the procedures followed by the supplier.
- *Order accuracy* – order accuracy is how closely shipments match customers' orders upon arrival, that is, having the right items in the correct quantities.
- *Order condition* – order condition is the lack of damage to orders due to handling.
- *Order quality* – order quality is how well the products work, that is, how well they conform to product specifications and customers' needs.
- *Order discrepancy handling* – order discrepancy handling is how well the firm addresses the discrepancies in orders after the orders arrive.
- *Timeliness* – timeliness is whether the orders arrive at customer locations at the promised time. The length of time between order placement and receipt is also included.

Customers' perceptions of a firm's logistics service quality begin to form as soon as they try to place their orders, and the perceptions continue to develop until the customers receive complete and accurate orders, in good condition, with all discrepancies addressed. Customers are also found to value each of the nine components differently (Mentzer et al. 2001). Firms will need to carefully develop their logistics strategy to cater for the varying needs of different customer segments.

MARKET ORIENTATION AND CUSTOMER SERVICES

In a JIT environment, any goal beyond delivering the right product to the right customer at the right time at the right price is a waste. From the customer perspective, waste is internal and external resources that are consumed without adding value to them, that is, if a customer is not willing to pay for the delivered products/services, then their existence is considered a waste. This means that the different types of wastes threaten many facets of the performance of the firm that customers may value. Hence their elimination has become an axiom.

To reduce wastes in customer services, a market orientation should not be neglected. It is concerned with organization-wide generation of market intelligence pertaining to current and future customer needs, dissemination of the intelligence across organizational functions, and organization-wide responsiveness to it (Kohli and Jaworski 1990). The organization-wide context of market orientation illustrates the importance of adopting a proactive attitude in doing business and staying close to customers.

The need to maintain a focus on customers and the market to serve is well recognized by many firms. The escalating expectations of customers have led many firms towards a market focus, orienting their activities to satisfying customers' needs and wants. As customer service is the centrepiece of business logistics, the customer definition of logistics is important to carry out the logistics activities. The key to success in business logistics lies in discerning and exploiting the relationships between customer definitions of logistics and the ways to translate them into real offerings. To this end, market orientation is important in enabling firms to understand the marketplace and develop appropriate logistics customer service strategies to better meet customer requirements (Lai 2003).

Viewed from this perspective, market orientation links customer requirements with relevant organizational functions within the firm. It helps to ensure that information collected from customers is used effectively as part of a customer service strategy, making customer perceptions and needs meaningful and explicit to organizational members. It also serves to communicate customer needs and requirements and the associated implications throughout the firm to ensure consistent decision making and actions, and to motivate corrective actions and methodological improvements when other functional areas fail to fulfil the needs of customers. The expected results of greater sharing of market intelligence throughout the firm are better coordination of the firm's efforts and that all functional areas work towards a common goal, that is, customer satisfaction. This is of particular importance in the development of customer service strategies to satisfy evolving customer requirements.

CUSTOMER SERVICE MANAGEMENT TOOLS

If a firm is going to elevate its customer service level in terms of increasing the availability of its products, this decision implies carrying higher inventory levels, renting more storage space, delivering bigger lots of inventories and incurring higher capital costs. Higher item availability inevitably affects the different elements of the logistics mix, as well as the total logistics costs. There are several tools that can help firms to determine their customer service levels.

Cost/revenue trade-offs – the sum of the expenditures on logistics activities such as transportation, order processing and inventory management can be considered as a firm's expenditure on customer service. Firms have to balance the trade-offs of the costs incurred from providing customer service and the levels of customer services offered to customers. The objective is to provide a firm with the lowest total logistics costs, given a specific customer service level. Firms should monitor their logistics costs, particularly the increasing logistics costs and the associated increases in sales, if the objective is to minimize the logistics costs given a specified level of customer service.

ABC analysis of customer service – ABC analysis (or the 80/20 principle) suggests that many situations are dominated by relatively few critical elements. For example, 80 per cent of a firm's sales might be generated by only 20 per cent of its customer base. Formulation of customer service policy using this method requires that companies focus on a few critical elements, for example, 20 per cent of their customer base. However, this approach fails to take into account the interest of potential customers.

Customer service audit – to reduce wastes in customer service, a firm needs to develop an index based on the important elements to its customers to monitor trends in order to maintain customer satisfaction. Firms should also conduct internal and external audits periodically to ensure that the internal and external dynamics are taken care of. The objectives of the audits are to identify the critical elements of customer service, find ways to control the performance of those elements and evaluate the quality and capability of the internal information system. Firms should ensure that the emphasis they place on any individual elements of customer service must be congruent with customer requirements. In addition, the customer service standards they employ should reflect what customers actually want, rather than management's perception of what customers want.

Importance-performance analysis – it can be used to identify the strengths and weaknesses of a firm's offering on the basis of consumers' perceived importance of, and the performance of, the various attributes delivered to them (Martilla and James 1977). In the analysis, customer perceptions of a firm on various performance measures are surveyed, as well as the importance they attach to these performance measures. The customer perceived performance measures are then classified into high/low categories and plotted on to a two-dimensional, four-quadrant importance-performance matrix (IPM) for interpretation. An example of an IPM is shown below. The vertical axis of the IPM indicates that the importance of the measures from low to high, and the horizontal axis represents their perceived performance from low to high. Positioning the vertical and horizontal axes of the IPM is a matter of judgment by the researcher, based on relative rather than absolute levels of importance and performance (Lai and Cheng 2003b). An example of IPM is shown in Figure 4.2. In the IPM there are four identifiable quadrants, namely: concentrate here (A); keep up the good work (B); low priority (C); and possible overkill (D). In quadrant A, performance in these measures is perceived to be relatively low while the measures are important to customers. This suggests that the firm should devote more attention to these measures – concentrate here. In quadrant B, these measures are considered to be very important to customers, while the firm delivers high levels of performance in these measures, suggesting that they should keep up the good work. In quadrant C, both the importance and performance levels of these measures are perceived to be relatively low, suggesting that the firm should put a low priority on improving these measures. In quadrant D, performance in these measures is perceived to be relatively high while the measures are unimportant to customers. This suggests that these overkilled performance areas have consumed excessive resources and the firms located within this quadrant should consider the reallocation of resources to other areas in need of strengthening.

Importance-performance matrix

	Low Performance	High Performance
High Importance	**QUADRANT A** Concentrate Here *High Importance* *Low Performance*	**QUADRANT B** Keep Up the Good Work *High Importance* *High Performance*
Low Importance	**QUADRANT C** Low Priority *Low Importance* *Low Performance*	**QUADRANT D** Possible Overkill *Low Importance* *High Performance*

Figure 4.2 Importance-performance matrix

Order Processing

The activities in order processing include customer placement of an order, transmission and receiving of the order, credit check, expedition of the order, distribution, and finally, the customer receiving the order and invoicing. The 'customer order cycle' refers to all the elapsed time between the points that a customer places an order to the point in time that the customer receives the order. One of the goals of managing the customer order cycle is to reduce the cycle time, which in turn helps improve customer service and logistical performance. An example of customer order cycle is shown in Figure 4.3. There are six stages in the customer order cycle, namely customer places order, order received by supplier, order processed, order picked and packed, order shipped to customer and order delivered to customer.

1. *Customer places order* – the entire order cycle starts at the moment when a customer places an order. As per conventional practices, firms often group orders in batches and handle the orders collectively at a later point in time. Grouping orders in batches allows firms to enjoy economies of scale in the transportation and administrative processes.
2. *Order received by supplier* – batches of orders will be sent to a supplier either on a regular basis or when the demand reaches pre-agreed volumes to enjoy economies of scale. The supplier will confirm with the firm via acknowledgment notices and start preparing the orders.
3. *Order processed* – the supplier will, according to the order details, process the necessary stock arrangements. In order to serve the customer's requests, another order cycle process may be invoked by the supplier to its own suppliers. The order information will also be forwarded to the accounting department for billing purposes.
4. *Order picked and packed* – items will be transferred from warehouses to the packing point. Protective materials or equipment will be used to handle products that require special care (for example, refrigerator for chilled products such as ice-cream and fresh fruits).

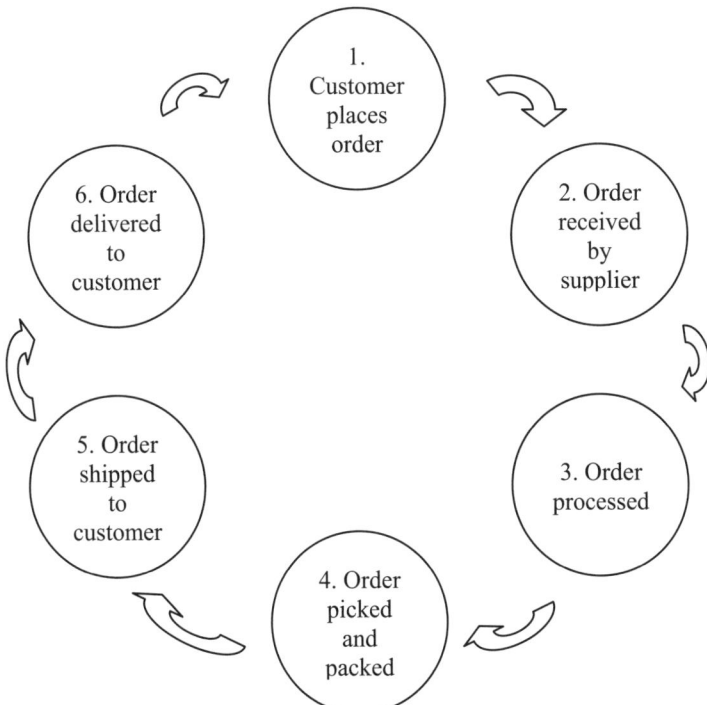

Figure 4.3 The customer order cycle

5. *Order shipped to customer* – once the ordered items are ready to be shipped, they will be transferred to the customer either by the supplier, the firm or a third-party carrier. Transportation mode will be carefully selected so that the orders arrive on time and the products are maintained in good shape.
6. *Order delivered to customer* – individual orders will be filled upon the arrival of the bulk shipment from the supplier.

Obtaining the necessary and correct information at the right moment is crucial for participants in a supply chain, so that they can adjust and adapt their own activities to the needs of the entire supply chain. Therefore, one of the goals of information sharing is to create the possibility for collecting and sharing useful and accurate data, thereby increasing transparency in the supply chain. The data that are usually shared include forecasts, historical data, market trends, production plans, as well as strategic information that affects partners in the supply chain. At the same time, there are exchanges of information about individual transactions between the firm and its vendors, including, for example, transferring of orders, order confirmation and invoices.

CYCLE TIME REDUCTION

In JIT logistics, the lead time of order management should be minimized. Lead-time reduction in a JIT environment will enable a firm to respond quickly to customer needs simply by reducing the time required to make the products/services available to customers. In recent years, such order processing activities as order placement, order confirmation and invoicing are increasingly handled electronically, for example, through EDI or over the Internet. In this way, the administrative tasks, as well as the transaction costs of the order cycle, are significantly reduced. Apart from using electronic means to shorten cycle time, other methods need to be deployed to eliminate waste because there are many areas where wastes can take place in order management.

- *Prolonged cycle time* – conventional order processing aims to satisfy the orders from a broader range of customers, for example, retailers, distributors and so on, and therefore requires a longer cycle time. The waiting time of a particular customer is therefore extended.
- *Batch processing* – processing orders from different clients collectively in batches is not compatible with the JIT philosophy because orders are not handled immediately upon a customer's request.
- *Administrative processing* – to process an order, firms will need to deal with its suppliers, carriers and banks. A transaction usually involves some type of formalities and paper-based administrative processes. These activities are wasteful because customers will not benefit from these procedures.
- *Order discrepancies* – shipping wrong items, misinterpreting product information or quantities and entering incorrect order details are examples of order discrepancies that should be eliminated.
- *Negotiations with suppliers* – if a firm has to negotiate with its suppliers to find the most competitive deal for every transaction, a considerable amount of time will be wasted as a result. Besides price negotiation, spending time on delivery and payment arrangements is also considered a waste of effort since these activities do not add any value to customers.

In JIT logistics, there is a vast amount of data transmission between suppliers and buyers involved throughout the order cycle. Order processing involves collecting, checking and transmitting order information among the various parties involved. Sales data need to be retained for market analysis and planning, cash flow forecasts and production. The order cycle duration will depend on the speed and accuracy with which sales information can be communicated through the order. Paper-based communication used to be the principal means for data transmission in the past. To make the process better, information such as production, supply and cycle time data will have to be continually shared with suppliers and customers. This sharing of information will become increasingly important as more and more companies adopt SCM practices that aim to link operations from suppliers' suppliers to customers' customers. Due to the advance in information technology, these once time-consuming tasks have now been greatly simplified.

Information technology is of significant value when we discuss the impact of JIT on order cycle as it helps to reduce communication delays. Information technology facilitates real-time communication and alleviates the extra transaction-related work that JIT brings about (as frequent processing of smaller batches will possibly invoke more transactions). The success of JIT is positively related to the utilization of electronic communications between firms and their suppliers (Wafa et al. 1996). Modern technology, especially advances in electronic communications, greatly 'speeds up' the order cycle (Teo et al. 2009). As will be discussed later, many enabling technologies have been developed to facilitate the flow of order information in an accurate and efficient manner. While information technology is valuable for reducing lead time in order cycle, firms should also pay attention to the electronic wastes created in the process. Electronic wastes refer to redundant or unnecessary data that is collected, managed and stored for no tactical or strategic reason. The amount of electronic wastes generated by a firm is typically great. It increases exponentially when one considers the data flows among members in a supply chain. Electronic wastes in order cycle are also a target for elimination under JIT logistics.

The lead time in the different stages of an order cycle can be efficiently managed through the implementation of JIT logistics in a supply chain. A supply chain will be capable of adjusting to changes and providing compensations between supply and demand, customers, plants and suppliers. Reducing the cumulative lead time or the sum of lead times for purchasing materials, manufacturing operations and product assembly allows a firm to reduce the planning horizon for production. Reducing the planning horizon allows a company to increase the accuracy of its demand predictions. Reducing inaccuracies in demand predictions diminishes the amount of buffer inventory that would otherwise be required.

The primary objective in making customer order cycle improvement is to shorten the cycle time. Bookbinder and Dilts (1989) proposed a JIT-based logistics information management model that requires firms to provide their suppliers with long-term order forecasts. Upon the establishment of long-term partnership programmes, firms supply their vendors with product-specific information including product numbers, projected product volumes, expected shipping dates and any anticipated future special delivery requirements. With medium- to long-term projections on hand, suppliers can better prepare themselves, in terms of workforce and equipment alignment, to serve instantaneous orders later on.

The order processing procedures in the order cycle under JIT logistics are essentially the same as before. Differences lie in *how* the orders are handled and *when* they are handled. A JIT order cycle is characterized by:

- immediate handling of orders;
- frequent transactions;
- increase in effort in handling transactions-related work.

Upon an order placement, JIT logistics urges firms to react instantaneously to an order, with minimal delays, by forwarding the requests to the respective suppliers, obtaining the designated product as soon as possible and delivering the product in the quickest manner to the hand of the customers. One should note that in a JIT context, several orders may be placed by firms daily. The order cycle is invoked even if the order quantity is relatively small.

How does JIT speed up the order cycle processes? A JIT order cycle goes hand in hand with an electronic communications network to transform paper-based procedures and manual operations into automatic processes, as seen in Figure 4.4. In an ideal setting, firms, suppliers, other intermediaries along the supply chain and even banks will be connected via an electronic network. This network enables real-time communication, as well as the transmission of order details among the chain partners. Once a customer places an order, a firm will forward the order to the supplier at once. A notification, from either the supplier or the firm, to the forwarder will be triggered at the same time. This order call will also raise an entry in the accounting systems of the firm, supplier and

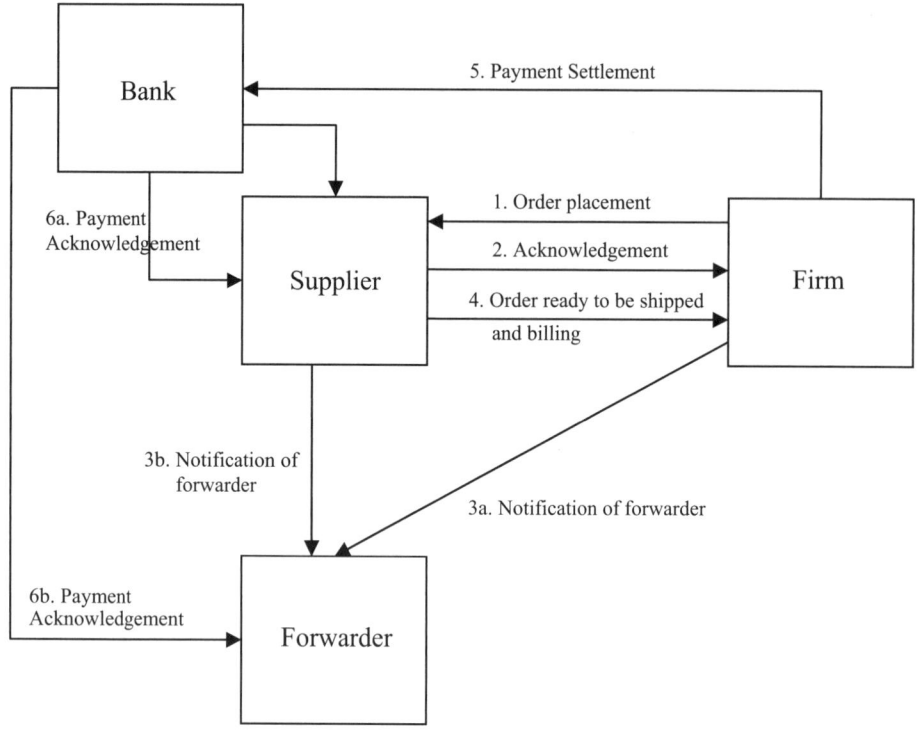

Figure 4.4 A JIT order cycle

forwarder to record the transaction. Once the product is ready to be shipped, billing notices from the supplier and the forwarder will be sent automatically to the firm, and the firm will advise the bank for payment settlement. The entire transaction process is conducted electronically so that order details are always dealt with at once. With the electronic network in place, orders are communicated to relevant parties on the supply chain within a matter of seconds.

PURCHASING AND COMPETITIVENESS

Purchasing represents the input side of logistics and provides a key opportunity for firms to enhance logistics performance and to create competitive advantages. It comprises the routine activities related to issuing purchase orders for needed products or services. Its role is to ensure uninterrupted flows of raw materials at the lowest total cost, improve the quality of the finished goods produced, optimize customer satisfaction and ensure continuous and reliable flows of goods into firms to meet their production and demand requirements. The purchasing activities may involve issuing purchase orders, contracting suppliers and checking the incoming purchased items. Purchasing links a firm's function with its many external suppliers. It emphasizes the need for continuity and stability in procurement, the development of key relationships with external suppliers and reliable, high-quality sources of supply.

The fundamental objective of purchasing is to provide the correct assortment of materials, parts or resale goods at the desired locations, when needed, and in a cost-effective manner. Traditionally, it has been the practice of purchasing to select suppliers on the basis of price or threaten suppliers to drop their prices, then simply buy products from whoever turning out to be the lowest-cost supplier. Recently, successful firms take a different approach that can best be described as partnership building. Firms rely on suppliers to improve quality, reduce costs and assist with product design and development. Such firms are learning to trust and develop their suppliers. In doing so, they are actively seeking better materials and reliable suppliers, working closely with and exploiting the expertise of strategic suppliers to improve the quality of raw materials, and involving suppliers and purchasing personnel in new product design and development efforts. This objective also requires that purchasing develop supplier performance capabilities. Such related activities have become just as important as purchasing. This means that suppliers have to improve their scheduling processes, reduce set-up time, reduce order-entry errors, change their facility layout and do whatever is necessary to improve their delivery performance.

The rationale is that sometimes firms find that their current suppliers are unable to support their JIT implementation and delivery requirements. They may need to seek other suppliers or work with suppliers to develop the skills needed to support their JIT efforts. Effectively managing and developing sources of supply often require purchasing to first identify suppliers who have the potential for excellent performance, and then approach these suppliers with the objective of developing closer relationships.

THE ROLE OF PURCHASING FOR JIT

The purchasing function in a JIT environment transgresses the traditional approaches employed in the past. The role of purchasing for JIT differs from the traditional approach in the nature of the relationship between a firm and its supplier, the frequency of deliveries and the number of suppliers a firm typically maintains. The element of purchasing that has been subject to the greatest impact is the quality of the supplier's products. There are many areas where wastes can be created from the purchasing activities. These wastes may involve:

- excessive raw materials;
- effort in handling defective materials;
- effort in dealing with a large number of suppliers, for example, price negotiation and delivery arrangements;
- idle time in waiting for deliveries;
- damaged materials due to inadequate packaging.

Traditional buyer-supplier relationships emphasize multiple sourcing, competitive bidding and the use of short-term contracts; these often adversarial relationships put the buyer against the supplier and focus primarily on the purchase price of the product instead of the capabilities of suppliers and how they can contribute to the long-term competitiveness of the buying organization. As mentioned, there has been a shift towards developing more long-term supplier relationships for cost and service advantages. Purchasing is an extremely important element in business logistics, since incoming materials' quality, delivery timing and purchase price are dependent on the buyer-supplier relationship and the capabilities of the suppliers (Lai et al. 2005a). Problems with suppliers will ultimately cause the end-product customers to get less and pay more.

One of the most critical issues in purchasing is supplier management. Simply put, this means getting a firm's suppliers to do what the firm wants, and there are a number of ways to do this. These involve assessing suppliers' current capabilities and then figuring out how to improve them. Thus, one of the key activities in supplier management is supplier evaluation, or determining the capabilities of suppliers. This occurs both when potential suppliers are being evaluated for a future purchase and when existing suppliers are periodically evaluated for performance purposes. A closely related activity is supplier certification. Certification programmes can either be company-designed and administered, or they can be internationally recognized and standardized programmes like the ISO 9000 series of certifications. Supplier certification allows buyers to assume that suppliers will meet product quality and service requirements, thus reducing duplicate testing and inspections, and the need for extensive supplier evaluation.

Over time, careful and effective supplier management efforts allow firms to selectively screen out poor-performing suppliers and build successful, trusting relationships with the remaining top-performing suppliers. These suppliers can provide tremendous benefits to the buyer firm and the entire supply chain. For instance, higher purchase volumes per supplier typically mean lower costs (causing a much greater impact on profits than a corresponding increase in sales), higher quality and better delivery service. These characteristics are viewed as strategically important to the firm because of their tremendous impact on the firm's cost and service advantages. Suppliers also see significant benefits from these relationships in terms of long-term, high-volume sales. These trading partner relationships have come to be termed strategic partnerships and are emphasized in JIT logistics.

To reduce wastes, the purchasing function can play an important role. For instance, it is not necessary to purchase in high volumes to enjoy economies of scale and maintain a certain level of stocks 'just-in-case'. The extra inventory incurs storing and carrying costs. If materials will arrive at the production facility when needed, firms will not need to 'stock up' in advance. Furthermore, firms should first carefully evaluate their suppliers on product quality and delivery reliability. Quality is emphasized not only to produce quality items but also to save the time and effort in dealing with defective materials.

Delivery reliability refers to whether suppliers are able to deliver the materials in good condition and on time to reduce unnecessary idle times. Delivery reliability is vital as any discrepancies in the delivery processes can easily make firms return to the practice of safety stocks. A consolidated supplier base frees up time and resources in dealing with different suppliers and encourages proactive efforts in organizing other value-generating activities such as improvement of distribution routes. The needs of the shops and production plants and so on will be pulled directly from suppliers to ensure that the supplies will arrive on time for continuous production. The Kanban system in JIT and other modes of ordering systems, for example, VMI, can be developed to serve this purpose.

The function of purchasing has manifested itself in the past as primarily concerned with obtaining the required parts at the lowest price. The shifting of focus towards quality represents but one change in the role purchasing plays in a JIT environment. Table 4.1 below presents a summary of these differences (Gonzalez-Benito 2002).

Table 4.1 Characteristic practices of JIT purchasing

Operational practices	Relational practices	Complementary practices	Quality practices
Use Kanban to coordinate with suppliers	Frequent communications	Involve suppliers in design and development	Select supplier selection based on quality and reliability
Frequent delivery (small batch sizing)	Single source (small numbers of suppliers)	Visit suppliers and provide suggestions for improvements	Monitor and control performance in quality and reliability
Reduction in inventory investment	Contracts in longer term	Supplier development such as providing training courses	Quality certification such as ISO
Exact delivery time (or tight time windows)	Repeated and lasting relationship		
Standardized containers	Relational programmes (initiatives to foster cooperation)		
Geographical concentration (through plants or warehouses)	Share benefits and risks		
EDI	Mutual assistance (through sharing cost information)		

The need to replenish inventory for production or distribution drives firms to purchase from suppliers on a continual basis. The waste reduction principle of JIT certainly imposes a solid guideline on a firm's purchasing practices. Firms must carefully decide on *when* supplies should be ordered and in what *quantity* to order. Numerous models and theories have been developed on purchasing for JIT. For instance, Kaynak and Hartley (2006) validated six factors essential for the implementation of JIT purchasing. These factors include top management commitment, employee relations, training, supplier quality management, transportation and quantities delivered. Although the past studies suggest different purchasing practices, most of them generally call for (Barnerjee and Kim 1995; Gunasekaran 1999; Benito et al. 2000):

- consolidation of the supplier base;
- having suppliers located in nearby areas;
- frequent deliveries in small quantities;
- having suppliers and firms establish long-term partnership or complementary relationships.

And, the studies all lead towards the goal of reducing:

- inventory carry costs;
- the need for inspection;
- re-work and late deliveries.

There are many benefits associated with embracing the JIT principles in purchasing; these include reduced administration costs, inventory levels and storage space, increase in product quality, and the identification and determination of quality problems. Despite the potential of these benefits, many supplier firms are reluctant to enter into a JIT relationship with their customers. Such a relationship requires suppliers to alter their practices and to adopt those commensurate with the JIT approach. The implementation requires alternations extending beyond merely the purchasing department. Nevertheless, there may be resistance from purchasing personnel to take part in JIT initiatives of a firm. The purchase resistance to JIT and the marketer response identified by Giunipero and O'Neal (1988) are summarized in Table 4.2.

SUPPLIER SELECTION

The difference in views between traditional and JIT approaches to logistics exist when dealing with suppliers. Single-source suppliers are preferred over multiple-source suppliers

Table 4.2   Purchase resistance and marketer response to JIT

Purchase resistance to JIT	Marketer response
The nature of our business is different	In general, JIT is applicable in most environments
As customers change our schedules, it is reasonable for us to change the schedule of our suppliers	To reduce the need for rescheduling, it is useful to develop a partnership relationship that gives forecast and schedule visibility to customers and suppliers
JIT seems like a supplier stocking programme, without benefits to our suppliers	A true JIT supplier will benefit from implementing the system in its own facilities. With this system in place it becomes possible to obtain long-term business, which permits improved planning and financial decisions
Our suppliers are too far away for the JIT initiative	Physical distance is not an issue for JIT, but its implementation requires better communication of schedules between buyers and sellers
It is the requirements by the auditors to seek frequent or annual bidding that dictates the selection of low price suppliers	It is useful to document savings achieved through better quality and reliable delivery that lowers the total cost
Successful JIT experiences are limited and not relevant to our operations	Interact with major professional groups in your area who can provide several success stories and the benefits associated with JIT

in a JIT environment because quality of both the product and the service is the most important criterion in vendor selection. In a JIT environment, suppliers are selected on the best price at a given level of quality as opposed to a low-cost bidder selection criterion. By reducing the number of suppliers and increasing the quantity of orders and the number of items purchased from each supplier, the buyer firm commits itself to a long-term relationship with its suppliers; improving communication (openness and frequency) between the buyer and supplier is an integral part of developing this relationship (Guinipero and O'Neal 1988). Buyers try to develop a nurturing long-term relationship with suppliers where the buyer and supplier work together to reduce the overall costs of doing business. Pine (1993) used the phrase 'long-term supplier interdependence' to represent this desired relationship. Pine (1993, p.124) stated that in this relationship '...companies work with their suppliers to achieve target costs and reductions over time and are willing to support them with engineering help, process innovations, and even extra time when necessary'.

On the other hand, Frazier et al. (1988) gives this description of the relationship between suppliers and buyers in a JIT environment: '...the need for close interfirm coordination in product development, quality assurance, and logistics is the key feature distinguishing JIT exchanges from other types of interfirm exchanges...' (Frazier et al. 1988, p.54). Part/material exchanges are designed in a creative, joint environment and continuous quality improvement is commonly a joint effort. The purchasing manager is responsible for developing suppliers' organizations and facilitating the implementation of JIT logistics that allow for an increase in suppliers' efficiency and flexibility.

Furthermore, Krause et al. (2001) developed a measure of competitive priorities for purchasing. The measure considers five areas of the competitive strategy of a firm, that is, cost, quality, delivery, flexibility and innovation, which are related to purchasing. A firm should consider all five areas in its supplier selection and retention decisions. The measurement items are listed in Table 4.3.

These competitive capabilities of a firm shape its ability to compete effectively in increasingly hostile environments. In particular, product innovation is fundamental for the continual prosperity of a firm (Koufteros et al. 2007a). The mere number of products and features that enter the market however would be inconsequential if their introduction fails to fulfil customer expectations for quality. With competitive capabilities being vitally important, firms constantly seek to engage in practices that would enhance their competitive capacity to innovate.

In order to enhance their product innovation capabilities and subsequently quality, many firms rely on their suppliers for product development contributions. Due to bounded rationality considerations, manufacturers focus on enhancing their own core competencies and depend on the complementary competencies that can be garnered from the involvement of their suppliers in product development and performance improvements, highlighting the strategic importance of supplier selection for purchasing decisions.

SUPPLIER MANAGEMENT

In addition to supplier selection, supplier management is critical for firms to be successful in practising JIT logistics. Traditionally the rivalry attitudes that suppliers and buyers hold towards each other imply passing risks on to each other and engaging in tough negotiations

Table 4.3   Competitive priorities and purchasing

Competitive strategy	Measurement Item
Quality	
Product reliability	The ability of a supplier to provide you with reliable inputs
Product durability	The ability of a supplier to provide your plant with durable products
Conformance to specifications	The ability of a supplier to conform to your specifications
Delivery	
Expediting	The ability, willingness and cost of a supplier to expedite a rush order
New product development	The amount of time it takes a supplier to develop a new part
JIT	The ability of a supplier to provide JIT delivery
Delivery speed	How quickly a supplier can deliver an order
Delivery reliability	The ability of a supplier to consistently deliver its products on promised due dates
Distance	Supplier location
Flexibility	
Volume flexibility	The ability and willingness and cost of a supplier to change order volumes
Mix flexibility	The ability, willingness and cost of a supplier to change the mix of ordered items
Modification flexibility	The ability of a supplier to design new products or make design changes in existing products
Cost	
Total cost	The total cost associated with the item including price, transportation, inspection and testing, cost of supplier non-conformance, customer returns and other associated costs
Cost information	The ability and willingness of a supplier to share cost data
Competitive pricing	The unit price of an item
Innovation	
Product innovation	The ability of a supplier to design new products or make changes in existing products
Technological capabilities	The level of technological capabilities the supplier possesses and is willing to use for your products
Technology sharing	The willingness of a supplier to share key technological information

to get the most at the lowest price from the other party. Supply management requires firms to work with their suppliers as partners and the boundaries between organizations become fluid as inter-organizational business processes are more integrated and information flows freely. In today's competitive business environment, price cut and superior customer service cannot be considered as long-lasting competitive edges. The rationale is that product innovations and price reductions are taking place daily. Through the building of better supplier relationships, both buyer and supplier firms involved in the logistics processes are working in harmony and have the same vision of driving down total costs and, most importantly, delivering superior products to customers. This requires supplier management, which calls for integrating the key business processes of buyer and supplier firms so that products, services and information that add value to customers and other stakeholders are provided. Table 4.4 compares the characteristics between market, relational and JIT exchange relationships.

Table 4.4 Comparisons of market, relational and JIT relationships on characteristics of exchange

Characteristics of exchange	Exchange relationships		
	Market	Relational	JIT
Time horizon of exchange	Short term	Moderate to long term	Long term
Focus of exchange	Price of core product	Focus on core product, with some attention to valued-added services	Equal emphasis on core product and value-added services
Number of inter-organizational linkages	Few	Moderate	A network of working relationships across functional areas
Frequency of communication	Low; mainly formal	Moderate; both formal and informal	High; both formal and informal
Nature of information sharing	Limited to economic transaction	Economic transaction with limited long-term planning	Joint product design and development and logistics related decisions with long-term planning
Frequency of shipments	Low and irregular	Moderate and regular	High, and subject to revision
Number of suppliers	Many	Moderate and selective for suppliers	Sole-sourcing and highly selective with few suppliers
Transaction costs	Low	Moderate	High
Asset specific investments	Low, if any at all	Moderately low	Moderate to high
Functional interdependence	Low and limited to delivery system	Moderate and involves only a few functional areas	Very high with extension to many functional areas
Risk level	Low	Moderate	High
Problem solving	Fire-fighting and reactive	Largely reactive	Proactive and oriented towards prevention

The logistics literature suggests that inter-firm coordination is positively associated with operational performance (for example, Stank et al. 1999). From the supply perspective, supplier development (for example, Scannell et al. 2000), supplier partnership (for example, Lai et al. 2005a), supplier involvement (for example, Vonderembse and Tracey 1999) and strategic sourcing (for example, Narahsimhan and Jayaram 1998) all positively influence the operation performance of the buying firm. One of the advantages of supplier management for JIT logistics is a firm's ability to decrease its supplier base to a more manageable size. Such action is expected to result in lower material overhead, fewer quality concerns, less inspection and other benefits derived from the potential of joint development projects utilizing the technology and technical expertise of suppliers. Firms are also finding it advantageous to involve suppliers in their internal decisions by establishing two-way customer/supplier EDIs, which expand the talent base, improve quality, reduce bottlenecks and increase cost reduction opportunities. Such systems also allow firms to involve suppliers in operations decisions and in material and design changes at an early stage, hence reducing the likelihood of scheduling problems.

In JIT, the purchased items should be delivered by suppliers when they are needed. A blanket purchase order or other forms of a basic agreement should cover the terms and conditions for the purchased items. Communication between the customer and supplier should be clear to ensure that supplies are delivered on time. Several techniques exist for controlling the flow of materials from suppliers to customers. Automatic inventory replenishment by a vendor is a technique by which a supplier determines the need for the required materials and makes frequent deliveries to a buyer firm's plant in an autonomous fashion. Also, the methods of written order for vendor delivery and verbal order for vendor delivery can be utilized.

SUPPLIER RELATIONSHIPS

The success of a supply chain depends not only on efficiency from optimizing resources, but also on the effectiveness of partner firms in carrying out mutually beneficial activities, that is, meeting customer requirements at the lowest possible cost (Lai et al. 2002). To achieve business success, it is imperative for firms to excel in the efficient and effective flows of goods and information in the supply chain. JIT usually requires major changes in supply management practices. Materials delivered in small lots of the required quantities must meet the quality and delivery specifications to avoid costly downtime (De Toni and Nassimbeni 2000). Supplier quality management practices include establishing long-term cooperative supplier relationships. With JIT, buyers usually reduce the number of direct suppliers used and move towards single sourcing (Cusumano and Takeishi 1991). It is easier to develop long-term, cooperative relationships with a smaller number of suppliers. Early involvement of suppliers in the buyer's product and service design improves quality and relationships.

Stanley and Wisner (2001) examined the association between the implementation of cooperative purchasing/supplier relationships, the quality of internal services and an organization's ability to provide quality products and services to its customers. Their study found support for the view that strengthening the relationship between the buyer and the supplier improves an organization's ability to deliver quality to customers. They suggested that managers should assess buyer-supplier relationships and take action where necessary to increase communication, solve problems and increase general awareness of the relationships between internal and external services and product quality. It involves agreements on specifications, exchanges of information, coordination and control between the buyer and supplier firms at the inter-organizational level that could affect the quality (conformance to requirements) of the delivery of the product or service, and the ability to achieve quality in the supply chain.

The importance of inter-organizational relationships to achieving cost and service advantages has been widely acknowledged (for example, Cannon and Homburg 2001). For instance, in his concept of lean supply, Lamming (1993) emphasized closer working relationships and transparent flows of information along the supply chain such that buyers can obtain the right quality of products/services at the right price, while suppliers can provide a quality supply profitably. Maloni and Benton (2000) tested a model on the influence of power in the supply chain. Their findings suggest that a stronger buyer-supplier relationship boosts performance throughout the supply chain. Fynes and Voss (2002) found that the buyer-supplier relationship has a moderating effect on quality practices and design quality. Their results suggest that it is desirable for suppliers to cultivate closer links with buyers to improve the quality of designs and related measures of quality performance. Recently, Lai et al. (2008a) examined the links between different buyer-supplier relationship variables and the contingent effect of business uncertainty on the links between these variables and supplier commitment. They found that expected relationship continuity mediates the effects of trust and relationship on commitment. Furthermore, the positive effect of trust on commitment was found to be stronger when business uncertainty is high than when business uncertainty is low. The above studies highlight the importance of managing supplier relationships for firms to improve performance.

A stable buyer-supplier relationship requires the involvement of firms beyond organizational boundaries to improve performance throughout the supply chain (Lai et al. 2005a). The stability of relationships goes beyond a simple, positive evaluation of the other party based on considerations of the current benefits and costs associated with the relationship. It implies the adoption of a long-term orientation towards the relationship – a willingness to make short-term sacrifices to realize the long-term benefits of the relationship (Dwyer et al. 1987). Such a long-term orientation needs stability to build confidence in the buyer-supplier relationship (Yang et al 2008). Therefore, for the buyer firm, the supplier's commitment to quality can be viewed as a long-term orientation in the buyer-supplier relationship. It requires a stable buyer-supplier relationship that will last long enough for the supplier firm to invest in a quality improvement system to meet the buyer's requirements and for both parties to realize the long-term benefits. Firms that form strong relationships with suppliers can better align their interests and goals with those of their suppliers (Lamming and Hampson 1996).

REVERSE LOGISTICS

Reverse logistics is a part of SCM that is also termed return management (Cooper et al. 1997), encompassing such traditional logistics activities as transportation and inventory management with a focus on getting back products from customers rather than moving products to customers (Mollenkopf and Closs 2005).

Reverse logistics is concerned with managing the flow of products/parts from the point of consumption back to the point of manufacturing for possible recycling, remanufacturing or disposal. Products are less likely to be disposed of unless all their reusable, remanufacturing or recycling potentials are fully exploited. Lower frequency of product disposal would mean higher economic benefits for firms and the environment (Dowlatshahi 2005). Reverse logistics can be considered as waste elimination and can be both cost effective and ecologically friendly by extending a product's normal life cycle beyond its traditional stage. The activities involved could encompass retailers, manufacturers and service entities. Recently, reverse logistics has received a lot of attention from operations managers as well as from many top executives due to its significant economic, environmental, managerial, regulatory and strategic implications for firms.

According to Flapper et al. (2005), the types of reverse logistics in a Closed-Loop Supply Chain (CLSC) can be classified as production-related, distribution-related, use-related and end-of life. The potential reasons why firms may choose, or be forced to practise reverse logistics are due to three main groups of business drivers: Profit, People and Plant.

Reversing the logistics is not as tidy as keeping it in forward motion. Items come in without packaging, as single pieces instead of cases or lots or pallets. The items for reverse logistics are certainly not scheduled. Unless the returned item is going to sit unnoticed in a warehouse forever, even something as simple as an exchange or repair will require at least some further supply chain involvement. For example, the store inventory must be updated to reflect the status of the returned item. Then, someone must be responsible for deciding what to do with the item – return it to inventory? Return it to the distributor, importer or manufacturer? Get it repaired? By whom? Where is it stored in the meantime? Who pays for the repair? Who pays for the return shipping? If a new order must be placed, what kind of priority does it receive and who notifies the customer when it arrives? Each of these seemingly routine decisions involves a policy and procedure that must be agreed on by the affected supply chain partners, or problems will surely arise.

One of the reasons that it is critical to track returns and have procedures in place to handle them is that they can point to bigger-picture issues, such as product defects or safety concerns. In the case of some products that contain hazardous materials (batteries, electronic equipment and so on), there are safe disposal requirements that must be adhered to and reported. A tracking system can also point to communication problems tied to customer expectations about the product, for example, if many customers return an item saying it 'doesn't work as well as I thought it would', this may require advertisements, marketing materials or product instructions to be modified to preempt future buyers' requirements.

Today's consumers take liberal policies for granted, and this adds some competitive pressure to the system. Customer service has become as important in the returns process as it was in the initial purchase of the item. If customers are requesting a replacement item, they want it and they want it now! A key component of the retailer's return plan must be to have sufficient capacity (staff, policies, inventory and so on) to make the transaction or complaint process as quick and hassle-free as possible – for the customer, that is.

Even other members of the supply chain that do not deal directly with consumers must plan for returns. They may be in the form of overstocks, end-of-season items, unsold or damaged goods, products that have been recalled and mountains of packaging materials. An attempt should be made to recover at least some of the items' original value, if possible. They may be repaired or reconditioned, resold to discounters, used for spare parts and so on. One consideration that is not directly related to customer satisfaction but important in terms of public perception is recycling of packaging materials and reusable parts. The general expectation is that companies will be good corporate citizens, which means having an environmental strategy to minimize waste.

GREEN SUPPLY CHAIN MANAGEMENT (GSCM)

Environmentally sustainable (Green) Supply Chain Management (GSCM) has emerged as an important organizational philosophy to achieve corporate profit and market share objectives by reducing environmental risks and impacts while improving ecological efficiency of these organizations and their partners (Zhu et al. 2008). As a synergistic joining of environmental and SCM, the competitive and global dimensions of these two topics cannot go unnoticed by organizations. For example, multinational enterprises have established global networks of suppliers to take advantage of country-industry-specific characteristics to build competitive advantage. Simultaneously, due to stricter regulations and increased community and consumer pressures, manufacturers need to effectively integrate environmental concerns into their regular practices and on to their strategic planning agenda. As a result, integrating environmental concerns into SCM has become increasingly important for business enterprises to gain and maintain competitive advantage. GSCM can be defined as integrating environmental concerns into the supply chain. Originally, GSCM was bound to purchasing issues and defined as integrating suppliers into environmental management processes (Walton et al. 1998). It consists of the purchasing function's involvement in activities that include reduction, recycling, reuse and the substitution of materials (Narasimhan and Carter 1998). As many industries have experienced increasing globalization and a shifting focus to competition among networks of companies, multinational enterprises need to establish global networks of suppliers that take advantage of country-industry-specific characteristics to build these competitive strengths and to balance efforts to reduce costs and innovate while maintaining good environmental (ecological) performance (Pagell et al. 2004). GSCM has emerged as an approach to balance these competitive requirements (Narasimhan and Carter 1998) and caused organizations to consider closing the supply chain loop (CLSC) (Seuring 2004). Within CLSC and GSCM practices, recoverable product environments, and the design of these products and materials, have become an increasingly important segment of the overall push in industry towards environmentally conscious manufacturing and logistics (Jayaraman et al. 1999).

According to Zhu et al. (2007), there are five types of GSCM practices including internal environmental management, green purchasing, customer cooperation with environmental concerns, investment recovery and eco-design dimensions. Internal environmental management is central to improving enterprises' environmental performance. It is generally believed that senior managers' support is necessary and, often, a key driver for successful adoption and implementation of most innovations, technology, programmes and activities (Hamel and Prahalad 1989). Cross-functional programmes encompassing GSCM and CLSC practices are not for the 'faint of heart' and require management's support for successful implementation. To ensure progress for environmental management, top management must be fully committed. Support also needs to exist from mid-level managers for successful implementation of environmental practices (Bowen et al. 2001). GSCM crosses all departmental boundaries within and between organizations, and this cooperation and communication are important to successful environmental practices (Apsan 2000).

External GSCM practices have become increasingly important for manufacturers. Zsidisin and Hendrick (1998) in a multinational investigation identified key factors for green purchasing including providing design specification to suppliers that include environmental requirements for purchased items, cooperation with suppliers for environmental objectives, environmental audits for supplier's internal management and suppliers' ISO14001 certification. Walton et al. (1998) put forward ten top environmental supplier evaluation criteria, among these, second-tier supplier environmentally friendly practice evaluation was viewed as the second most important criterion. In addition, large customers have exerted pressure on their suppliers for better environmental performance, which results in greater motivation for suppliers to cooperate with customers for environmental objectives. Customer cooperation and green purchasing concern types of customer relationships that occur in CLSC. Without green purchasing and customer cooperation practices, product take-back and other product reintroduction markets may not become as developed (van Hoek 1999).

US and European enterprises have considered investment recovery as a critical aspect of GSCM (Zsidisin and Hendrick 1998) and CLSC (Thierry et al. 1995). Investment recovery typically occurs at the back end of the supply chain cycle or as a method to 'close the loop'. Investment recovery concerns the different types of recovery processes that occur in CLSC. Toffel (2004) stated that concerns for the end-of-life products are motivated by legislation across Europe. Even in non-regulated markets, some manufacturers have engaged in product recovery to reduce production costs, enhance brand image, meet changing customer expectations, protect aftermarkets and pre-empt pending legislation or regulations. Sale of excess inventories and capital equipment are also aspects of investment recovery. As an example, in the US, many firms, as part of their reverse logistics operations, have started selling unwanted products in online auctions (Tibben-Lembke 2004). The other set of practices defining the 'back end' of the supply chain includes the relationship with customers on environmental issues. In many cases, international companies are now emphasizing the need for their suppliers to maintain an environmentally benign position so that their products are not boycotted for environmental reasons.

Most of the environmental influence of any product or material is 'locked' into the product at the design stage, when materials and processes are selected and product environmental performance is largely determined. Pioneering firms have learned that making product returns profitable relies on good design (Krikke et al. 2004). It has been argued that for effective product stewardship and reverse logistics practices, eco-design (which would include design for disassembly, design for recycling and design for other reverse logistics practices) is necessary (van Hoek 1999). Thus, eco-design or Design for Environment (DfE) is an important and emerging GSCM practice to improve companies' CLSC. Eco-design concerns types of product-oriented relationships that occur in CLSC. It is meant to address product functionality while simultaneously minimizing life-cycle environmental impacts. The success of eco-design requires internal cross-functional cooperation within the company and external cooperation with other partners throughout the supply chain.

It should be noted that greening the supply chain is increasingly a concern for many business enterprises and a challenge for logistics management in the twenty-first century. Of particular concern is how to arouse their environmental awareness and put it into practice in the logistics activities of their supply chains. GSCM represents a recent and important inter- and intra-organizational set of environmental management practices useful for logistics management in this context. It is designed to incorporate environmental considerations into decision making at each inbound logistics stage of material management all the way through to the outbound logistics stage of post-consumer disposal. Governmental and other public pressures will accelerate the process of firms to implement environmental practices, especially practices within GSCM in the years ahead.

ENABLING TECHNOLOGY

Porter (1985, p.50) suggested that gaining competitive advantage requires exploiting a system's interdependencies; this '...usually requires information or information flows that allow optimization or coordination to take place'. For example, Womack and Jones (2003) described a potential future scenario of using information technology to integrate the distribution system with the production system in a Japanese automobile system. A database would be developed for each consumer that includes specific information about each buyer's household. Then, a computer, upon request from the buyer, is used to recommend the appropriate automobile based on the latest buyer information data. This computer system could also provide guidance for financing, insurance and so on. Other examples were used to show how customers make input directly into the design of the product they purchase. The customer, through CAD, changes aspects of the design of their car or house. The CAD data are then used for product design documentation, order entry, processing and so on.

In the purchasing process, data communications between buyers and suppliers are numerous including the items and quantities ordered, prices, delivery dates, delivery addresses, billing addresses and payment terms. A challenge for purchasing is to ensure that the data communications are carried out in a timely and accurate manner. The effectiveness of a supply chain depends on several factors. One important factor is the use of technology to conduct business transactions and to facilitate the sharing of information (within the company, as well as outside the company) and collaboration with suppliers and customers (Singh et al. 2007). Information technology can be used to complement order processing for JIT logistics by providing for improved information flows throughout the supply chain and allowing for a more comprehensive analysis of the expanded data. A reliable communication infrastructure paves the way for timely and efficient information exchange among partners (Lai et al. 2005b). For example, using EDI technology, manufacturers can provide up-to-the-minute information about their production needs by giving suppliers access to their production planning and control systems, and suppliers can arrange deliveries without the need of any paper transactions.

Similarly, timely payments can be arranged using EDI. Reduction of payment delays lowers the cost of doing business significantly, makes supply chains more efficient, and gives users a competitive advantage. The integration of many IT-enabled electronic commerce tools – bar coding, electronic messaging, EDI, global network management and the Internet – allows supply chain partners to attain significant productivity gains. The fruits of information integration, such as reduced cycle time from order to delivery, increased visibility of transactions, better tracing and tracking, reduced transaction costs and enhanced customer service yield greater competitive advantages for all participants in the supply chain. Today, modern logistics management relies heavily on access to accurate and timely information (Ngai et al. 2008). They need to have enabling technologies to obtain the related information for effective decision making. These different major enabling technologies are elaborated below:

Electronic data interchange exchange standards

EDI represents a company-to-company data exchange mechanism in the form of electronic transfer. Data are transformed into pre-agreed formats and transferred between trading partners. Typical business forms, such as invoices, bills of lading, purchase orders, shipment schedules and production schedules are transferred via EDI (Clouse and Gupta 1990).

The immediate benefits of EDI include the elimination of labour and material costs associated with printing, mailing and dealing with paper-based transactions, and the reduction of manual data-entries. Other derived benefits are shortened order cycle, decreased labour, freight, and material costs, availability of more complete, timely and accurate information (Clouse and Gupta 1990).

VAN

EDI provides a means for business partners to communicate transaction details instantly. The transaction details are designed as a 'transaction set.' Each industry has it own form of transaction set. A problem immediately arises: how do people from different industries communicate if they do not speak the same language? VAN provides a solution to this problem. It is a common interface between any sending and receiving systems and helps to manage transactions, translate communication standards and reduce the number of communication linkages (Bowersox et al. 2002).

Internet

Until just a few years ago, EDI was the most widely accepted platform for the transmission of online information about transactions in a standardized format between companies, for example, purchasing orders, invoices, order confirmations, payment transfers, transport bookings and inventory status. The implementation of EDI connection is, however, relatively costly and assumes that the transmission of large amounts of standardized transaction is profitable. EDI solutions are tailor-made, though, to specific business relationships. As a result, the initial investment is high, and is lost completely if the business relationship dissolves. An additional factor is that EDI can only transfer statistical transaction data, not real-time information such as actual production, actual inventory or actual demand. Finally, EDI has a hierarchical structure with a one-to-several architecture, typically involving one dominant company in the supply chain, which develops EDI connections to its strategic partners. In smaller companies, EDI would normally only be used if it is required by important customers. Within the grocery industry, where there are large quantities of daily transaction data, EDI is widely used.

The Internet has made available a business platform that combines the best of the earlier existing platforms. Like EDI, the Internet facilitates structured communication. At the same time, barriers to access are low, creating no hindrances to small companies using the Internet as a business platform. The prices of software programs that can integrate external data transfer with a company's back-office solution are falling drastically. Data transmission via the Internet is increasingly standardized as XML-files (eXtensible Markup Language). The Internet can, unlike EDI, support a many-to-many configuration of a supply network.

The widespread availability and use of the Internet offers firms opportunities that did not exist before. These opportunities are made possible because it is now so easy and relatively inexpensive for firms to connect to the Internet. Once connected, firms can send data to and receive data from other firms that they do business with regardless of the particular computers or software that individual firms may be using to run their internal operations. Based on this data sharing, opportunities exist to achieve tremendous logistics efficiencies and significant increases in customer service and responsiveness. There is also a trend in the business world today to use electronic procurement, which is about handling purchasing functions online using software applications that are Internet-based. In doing so, suppliers can upload their catalogues, order forms and other data on a website that potential buyers can access and place orders on. In business-to-business applications, the transactions are secure and can be tracked by the involved parties.

Electronic logistics

The Internet is an enabler of electronic logistics. Electronic logistics is basically the use of information and communication technologies both internally and externally in a supply chain, with the goal of ensuring online, real-time information exchange between parties. Electronic logistics makes extensive use of the Internet, but intranet, extranet, bar code systems and electronic media are utilized as well.

Electronic logistics is an extension of the logistics concept including, on the technical side, the business processes between parties, which to the greatest possible extent occur via electronic networks. Sharing of information and applications are also important characteristics of electronic logistics. The most important changes occur, however, on the business side, where information technology improves communication between parties, allowing shared planning and control, while enabling direct interaction with customers and discerning their individual needs and wishes.

From the coordination-theoretic perspective, which is a body of principles about how business activities can be coordinated amongst multiple organizations that work together towards common goals such as cost reduction, logistics management is increasingly recognized as an important area for innovation and investment in IT. This is because information is an element that holds firms in a logistics chain together in response to ever changing market requirements (Lai et al. 2008b). By establishing information-sharing linkages, the application of IT facilitates logistics operations as it provides electronic connections among members of a logistics chain. The application of IT can contribute to the management of logistics in a number of ways from information storage to active, dynamic and interactive systems that support the performance of different logistics activities. Electronic logistics can add value to stakeholders including customers and suppliers by furnishing an electronic connection for trading convenience and information exchange that could differentiate the customer service and reduce the operations costs of a logistics chain. Such electronic connections are highly desirable for logistics management, which requires the establishment of inter-organizational information networks and the construction of an integrated Logistics Information System (LIS) (Ngai et al. 2008).

A number of firms have already incorporated an electronic element in support of their business activities, for example, in the forms of order transfers via EDI/ Internet, purchasing through electronic marketplaces or Internet sales to customers. However, few firms base their work on an electronic platform and have integrated their information and planning systems with the different links in their supply chains, from customers to vendors. Electronic logistics connects individual firms with their networks of suppliers, transporters and customers. The point is to utilize shared forecast data, production plans, customer orders and inventory status – in short data that influence the total supply chain's efficiency and flexibility. The goal is to align supply and demand. Transparency throughout the logistics processes, based on real-time information, is a pre-requisite for firms to be able to act proactively rather than reactively in relation to the links closest to them in their supply chains.

Inventory Management

In a JIT environment, inventory is considered as wasteful, excessive and inherently evil and the aim of the firm should not be just to reduce it but to completely weed them out (Schonberger 1982a). Over the past decades, there have been many calls for a revolution in inventory policies of firms. There has been a belief that firms with lean inventories are more valuable than firms with bloated inventories and the financial markets would reward firms that cut inventories and punish those that did not do so. In a study on the inventories of US companies between 1981 and 2000, it was found that firms with abnormally high inventories had abnormally poor stock returns. Firms with abnormally low inventories had ordinary stock returns. Firms with slightly lower than average inventories performed best over time. They outperformed average firms by about 4.5 per cent per year on average (Cheng and Wu 2005). Their findings suggest that inventory management is important for the performance of firms from the perspective of stock market valuation.

It is generally recognized that the implementation of JIT will result in significant reduction in inventory. Inventory levels are a key measurement of the success of JIT logistics. The rationale for the emphasis of JIT logistics on inventory management is simple to elaborate. To strive for a level of zero inventories, produce items at a rate required by the customer, eliminate all unnecessary lead times, reduce set-up costs to achieve the smallest economical lot size, optimize material flow from suppliers through the production process to Point-of-Sale (POS) of the finished product, so that inventories are minimized, high-quality JIT delivery from suppliers are ensured, safety stocks are minimized and a Total Quality Control (TQC) programme is implemented that minimizes scrap, re-work and resultant delays in production and distribution.

JIT logistics emphasizes the coordination of the logistics activities of the firms in a value chain. Pulling materials through the supply chain in response to demand patterns

On the other hand, Alles et al. (2000) found that inventory levels drive quality. Lower inventory levels force workers to work smarter (that is, supply more knowledge effort) as the absence of inventories forces workers to immediately trace the causes of any defects. The instantaneous, context-specific feedback and identification of root causes improves the workers' ability to solve critical manufacturing problems and fine-tune the production process.

THE BULLWHIP EFFECT

One of the major causes to wastes in inventory management is information distortion in the logistics process. In a traditional supply chain, information about the final customer's actual demand is often delayed from one link in the supply chain to the next. This phenomenon is often called the bullwhip effect (Lee et al. 1997). What happens is that small changes in product demand by the consumer at the front of the supply chain translate into wider and wider swings in demand experienced by firms further back in the supply chain. Firms at different stages in the supply chain come to have very different pictures of market demand and the result is a breakdown in supply chain coordination. Firms behave in ways that at first create product shortages and then lead to an excess of products.

To illustrate, each link in a supply chain will often have its own system for distributing and organizing activities based on the information it receives from the subsequent link in the supply chain. The vendor will purchase and produce on the basis of the orders received from the producer. The producer will arrange its activities depending on the orders received from the wholesaler. The wholesaler will respond as necessary based on the orders from the retailer. The retailer will place its orders based on the behavioural demand patterns of the final customer. All links in the supply chain will have their own built-in ordering routines, which are based on reorder points and quantities. This practice creates larger fluctuations in the demand the further upwards the supply chain one comes. As a result, there is an increase in variability of orders up the supply chain. This bullwhip effect is illustrated in Figure 4.5.

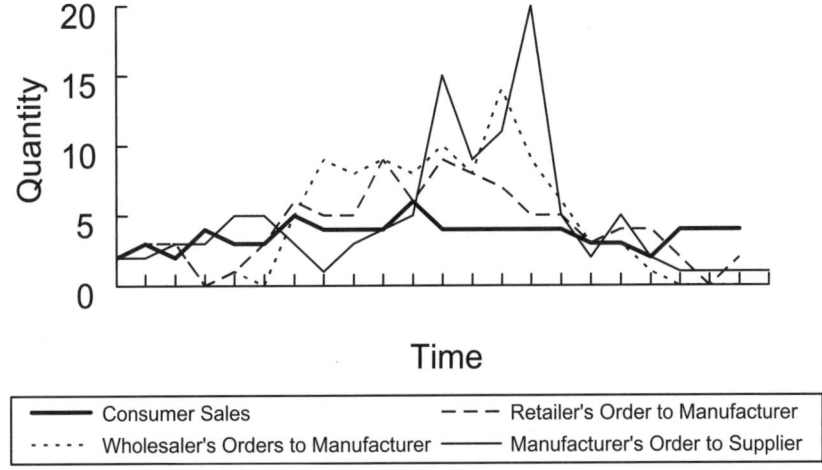

Figure 4.5 The bullwhip effect in the supply chain

Even though the end-item demand may be relatively constant, forecasts and their corresponding orders as one moves up the supply chain can become amplified, causing the bullwhip effect. These variations in demand cause problems with capacity planning, inventory control, and workforce and production scheduling and, ultimately, result in lower levels of customer service and higher total supply chain costs. According to Lee et al. (1997), there are several major causes:

Demand forecast updating

Demand forecasting based on orders received instead of end user demand data will inherently become more and more inaccurate in moving up the supply chain. Firms that are removed from contact with the end user can lose touch with the actual market demand if they view their role as simply filling the orders placed with them by their immediate customers. Whenever a buying firm places an order, the selling firm uses that information as a predictor of future demand. Based on the information, sellers update their demand forecasts and the corresponding orders placed with their suppliers. As lead times grow between orders placed and deliveries, safety stocks also grow and are included in any orders as they pass up the supply chains. Thus, fluctuations are magnified as orders vary from period to period and as the review periods change, causing frequent demand forecast updating. One way to counteract this distortion in demand forecast is for all the firms in a supply chain to share a common set of demand data from which to do their forecasting. The most accurate source of this demand data is the supply chain member closest to the end use customer (if not the end use customers themselves). Sharing POS data among all the firms in a supply chain goes a long way towards taming the bullwhip effect because it lets everyone respond to actual market demand instead of supply chain distortions.

Order batching

Order batching occurs because firms place orders periodically for different amounts of the product to minimize their order processing and transportation costs. In a typical buyer-supplier scenario, demand draws down inventories until a reorder point is reached at which the buyer places an order with the supplier. Inventory levels, safety stocks and the desire to order full truckloads or container loads of materials may cause orders to be placed monthly or even less often, or at varying time intervals. Thus, at one point in time, the supplier gets an order of some magnitude; then, until the next order is placed, there is no demand at all for the supplier's goods. This type of order batching amplifies demand variability and adds to the bullwhip effect. Another type of order batching can occur when salespeople need to fill end-of-quarter or end-of-year sales quotas, or when buyers desire to finish year-end budget allotments. Salespeople may generate production orders to fill future demand, and buyers may make excess purchases to spend the budget money. These erratic, periodic surges in consumption and production also increase the bullwhip effect. If the timing of these surges is the same for many of the firm's customers, the bullwhip effect can be severe. One way to address demand distortion caused by order batching is to reduce the cost of order processing and transportation. This will get lot sizes smaller and orders to be placed more frequently. The result will be a smoother flow of orders that distributors and manufacturers will be able to handle more efficiently. Ordering costs can be reduced by using electronic ordering technology. Transportation costs can be reduced by using Logistics Service Providers (LSP) to pick up many small shipments from suppliers and deliver small orders to many customers in a cost-effective manner (Lai 2004).

Product pricing

Product pricing causes product prices to fluctuate, resulting in distortions of product demand. If special sales are offered and product prices are lowered, it will induce customers to buy more products or to buy products sooner than they otherwise would (that is, forward buying). When suppliers have special promotions, quality discounts or other special pricing discounts, these price fluctuations result in significant forward buying activities on the part of buyers, who are stocking up to take advantage of the low price offers. Forward buying occurs between retailers and consumers, between distributors and retailers, and between manufacturers and distributors, due to pricing promotions at each stage of a supply chain, all contributing to erratic buying patterns and, consequently, the bullwhip effect. If these price discounts become commonplace, firms will stop buying when prices are undiscounted and buy only when the discount prices are offered, contributing even further to the bullwhip effect. To deal with these surges in demand, manufacturers may have to vary capacity by scheduling overtime and undertime employees, finding places to store stockpiles of inventory, paying more for transportation and dealing with higher levels of inventory damage as inventories are held for longer periods. Answers to this problem generally involve invoking the concept of 'everyday low prices'. If the end customers for a product believe that they will get a good price whenever they purchase the product, they will make purchases based on real need and not other considerations. This in turn makes demand easier to forecast and firms in a supply chain can respond more efficiently.

Product rationing

This is the response that manufacturers take when they are faced with more demand than they can meet. The problem occurs when demand exceeds a supplier's finished goods available, and in this case, the supplier may allocate products in proportion to what buyers ordered. Thus, if the supply of goods is 75 per cent of the total demand, buyers would be allocated 75 per cent of what they ordered. When buyers figure out the relationship between their orders and what is supplied, they tend to inflate their orders to satisfy their real needs. This strategy is known as shortage gaming. Of course, this further exacerbates the supply problem, as the supplier and, in turn, their suppliers struggle to keep up with these higher demand levels. When, on the other hand, production capacity eventually equals demand and orders are filled completely, demand suddenly drops to less-than-realistic levels, as the buying firms try to unload their excess inventories. There are several ways to respond to this. Manufacturers can base their rationing decisions on the historical ordering patterns of a given distributor or retailer and not on their present order sizes. This eliminates much of the motivation for shortage gaming that otherwise occurs. Manufacturers and distributors can also alert their customers in advance if they see demand outstripping supply. In this way product shortages will not take buyers by surprise and there will be less panic buying.

The bullwhip effect can have causes other than insufficient information between links in a supply chain, including:

- shipment consolidation, in order to achieve better distribution capacity utilization;
- order consolidation, in order to attain quantity discounts or to simplify ordering;
- sales campaigns, which can encourage customers to purchase in larger quantities than they actually need. This tendency can give an incorrect impression of demand;
- speculation in shortage situations, which leads customers to order more than they need in the hope of securing a large percentage of the allocated products;
- delays as a result of administrative routines, for example, weekly ordering or long transport time (for example, container transport from China to the US).

In addition, the increase in the globalization of logistics activities and the practice of subcontracting manufacturing and offshore sourcing may extend the lead time required to complete product flows in the whole process and therefore attenuate the bullwhip effect. The extended lead time will cause an increased risk for members in a supply chain to lose visibility and control of the logistics activities in other parts of the chain, for example, inventory record, delivery status, actual demand and forecasts and production plans, and lead to wastes in inventory management. Christopher and Lee (2004) argued that inventory wastes in the supply chain are due to a lack of confidence in the involved parties, which reflects their perceptions of performance reliability at each step in the chain. Performance reliability is concerned with how much faith the various parties in a supply chain have in the ability of those in the 'upstream' and 'downstream' to do what they say they are going to do. Without supply chain confidence, buffering is a popular means employed by logistics managers to hedge against demand and supply uncertainties in the supply chain. Another popular means used by logistics managers to hedge against demand and supply uncertainties is to invest in excessive capacity.

According to Christopher and Lee (2004), greater confidence of firms in a supply chain will increase their willingness to reduce safety stock and substitute information for inventory. By exchanging information on forecasts, production plans, inventory and principles of allocation, as well as by integrating information systems between the involved parties, the unfortunate consequences of large fluctuations in demand can be reduced considerably. This is because shared information reduces uncertainty and thus reduces the need for buffer stock. This would enable firms to be more responsive and the supply chain to become demand driven rather than forecast driven (Cheng and Wu 2005; Wu and Cheng 2008). Information transparency in the logistics process would greatly reduce wastes in inventory management.

AGILITY AND LEAN THINKING

One inventory challenge in logistics concerns the need for flexibility to meet the increasingly segmented needs of downstream customers, particularly in regard to product preferences and quantities demanded. A solution to this challenge is agility, which confers the ability to efficiently change operating states in response to uncertain and changing market conditions (Nagel and Bhargava 1994). Agility involves flexibilities of several sorts, and includes the capability to do unplanned, new activities in response to unforeseen shifts in market demands or unique customer requests (Prince and Kay 2003). Agility requires the use of market knowledge and a virtual corporation to benefit from rapidly changing market opportunities. It emphasizes flexible, timely action in response to rapidly changing demand requirements (Christopher and Towill 2002). According to Goldsby and Garcia-Dastugue (2003), an agile approach is preferred when product variety is large, demand is highly unpredictable and product cycles are short. As an agile manufacturer, Dell Computer utilizes a rapid configuration system for its custom-built computer products. This system drives the company's efficient consumer-direct marketing strategy. After first collecting in-depth information about customer requirements, Dell then regularly transmits the information to suppliers. Further, Dell's practice of making its suppliers in-house partners achieves tight internal coordination without excess personnel or costly inventory.

Another inventory challenge is concerned with materials rationalization, where many products have unnecessary or excessive materials built into them. A firm's ability to achieve material reductions provides it with a competitive advantage over rivals that possess less effective control techniques. The lean thinking advocated by Womack and Jones (2003) suggests that relentless elimination of waste is one solution to the excess material problem, an idea best exemplified by the lean manufacturing philosophy of the Toyota Production System, which is congruent with the JIT philosophy in this book. Activities that consume resources but generate no redeeming value in the eyes of customers are wastes that must be eliminated in the 'lean' paradigm. In comparison with a mass production approach, a lean company calls for less inventory and incurs fewer defects while providing greater variety in products.

There are hybrids of the agile and lean approaches, termed 'leagile'. Christopher and Towill (2002) identified three distinct hybrids. The first hybrid approach embraces the Pareto (80/20) principle, recognizing that 80 per cent of a company's revenue is generated from 20 per cent of its products. It is advisable that the fast-moving products that make up the dominant 20 per cent of the product line can be produced in a lean, make-to-stock manner, given that demand is relatively stable for these items and that efficient replenishment is the appropriate objective. On the other hand, the remaining 80 per cent should be produced in an agile, less anticipatory manner, perhaps even employing make-to-order production to generate supply for items only when they are ordered. It is common for manufacturing facilities to be designed so that some lines are designated for efficient processing of fast-moving products while others are designated as small-batch lines with quick, frequent changeovers in support of the slower-moving items.

Another hybrid involves having temporary capacity to meet the needs of peak demand. Most companies experience a base level of demand over the course of the year. This base level of demand can be accommodated in a lean manner, using the company's own resources to employ smooth production principles to maintain highly efficient operations. However, when demand spikes over the course of peak seasons or heavy promotion periods, outside capacity is procured to meet the heightened demands of these distinct windows. The procurement of outside capacity for coverage in these situations is viewed as the agile component of this hybrid approach. Many companies engage leagile supply, manufacturing, and logistics to support seasonal demand. The third hybrid approach is termed postponement, which is about delaying the final form of the product until an order is received from customers dictating the quantity and qualities of the goods demanded (Feitzinger and Lee 1997). This approach works best when goods can be developed from common materials into a near-finished state with final touches to the product providing for a diverse assortment that accommodates distinct customer needs. The premise calls for lean operations in the production of generic, semi-finished products, and agile accommodation in the customization process. It is on the basis of accommodating diverse needs efficiently that one may refer to such an approach as 'mass customization' (Feitzinger and Lee 1997).

INVENTORY MANAGEMENT TOOLS

The ideas behind JIT logistics are to connect buyers and suppliers in the logistics activities throughout an entire supply chain to jointly bring about total supply chain cost reduction and service improvement. This is accomplished through ensuring continuous process improvements, eliminating NVA practices, matching inventory flows with consumption and linking information flows to inventory flows. Anything that disrupts the logistical flows of goods in the supply chain creates systemic costs in terms of time and inventory. There exist several useful management tools that can aid management to better manage their inventories. These include:

Kanban

This is a restocking system that has evolved through the refinement of the production system at Toyota Motor Corporation. This management tool for JIT logistics helps to control the delivery of the necessary quantities and timing of the various parts throughout the operations among suppliers with which the focal firm interacts. Kanban is an information system or card system used to help 'pull' the necessary parts into each operation where customer demand creates the initial pull of the system. Kanban is analogous to a travel card, floating order card or travelling requisition as utilized in the purchasing function (Chase et al. 2006). The movement of materials through each of the activities is comparable to the movement of a water bucket passing in an old-fashioned fire-brigade. To provide the system responsiveness and product availability necessary in an increasingly unpredictable environment, the total system must use Kanban throughout the pipeline of materials flow. Implementation of Kanban piecemeal, starting with customer demand and moving upstream in the pipeline of material flow provides the tying mechanism for the continuous transmission of information in processing and across all activities. This allows for real-time decision making.

Materials requirement planning (MRP) I

Materials Requirement Planning (MRP) I is an expensive computerized system used by an organization to plan and control the materials and resources needed for the production of goods. The function of MRP I is to calculate the requirements from materials through the application of bills of materials, inventory status information and a master production schedule. It is a viable technique when the demand of an item is derived from the demand of another item. It is beneficial to apply MRP techniques when:

- the usage or demand for materials is discontinuous or volatile during a firm's regular business cycle;
- the demand for certain supplies depends on the production of other inventory items or finished products;
- the purchasing and manufacturing units, as well as the suppliers, are flexible in handling order replacement.

Manufacturing resources planning (MRP) II

Manufacturing Resources Planning (MRP) II is an advanced model of MRP I that further integrates a firm's financial, marketing and purchasing functions in the planning and control of its production operations. The objective of MRP II is to plan resources for all the production activities, which include materials, labour and production capacity. The execution of this function is practicable through the application of bills of materials, master production scheduling and MRP I.

Distribution requirement planning (DRP)

Different from MRP that is production-oriented, Distribution Requirement Planning (DRP) starts with customer demand and works backwards to derive a system-wide plan for ordering finished products, components and materials, respectively, as well as a time-based plan for distributing products from plants and warehouses to POS for customers. DRP demonstrates excellent customer responsiveness since it can adjust and readjust its ordering patterns to accommodate dynamic and changing inventory needs.

JIT II

This builds on JIT by physically placing a supplier's employee(s) at a customer's site. The supplier's representative – referred to as an inplant – has the authority to conduct transactions between the supplier and customer, attend customer's engineering and design meetings to help plan product development and refinement, gather marketing and sales information from the customer, and determine the customer's present and future needs. The inplant reports to the supplier to adjust production and inventory as quickly and cost efficiently as possible.

Flexible manufacturing

This is the alignment of suppliers, equipment, processes and other resources in a bid to adapt to the changes in the shortest time and at the lowest costs. It is critical for firms to be able to adjust their operational settings to hedge against uncertainties in demand.

ABC analysis

This is special technique used to classify inventory according to their importance. The classification mechanism follows the Pareto's Law, or the '80-20' rule, which, in the inventory management context, states that a relatively small number of items or stock-keeping units may account for a considerable impact or value. Inventory can be classified into three groups, namely those with the greatest impact constitute group A, while those with lesser impact form groups B and C, respectively. The baskets of items in the three groups are essentially different for different firms. To perform ABC analysis, firms will first identify their own set of criteria, then rank and classify the items accordingly (Coyle et al. 2003).

Benchmarking

The goal of benchmarking is to learn and enquire about process and product innovations that are proven to be successful in other organizations (Bagchi 1997). Firms benefit from benchmarking through a systematic process of learning and adapting the best business practices. Benchmarking helps firms identify the best inventory management practices in their respective industry. The best practices can be used for developing sources of distinctive competence and uncovering weaknesses that need to be addressed. The idea is to align the operational activities on the functional level with the strategy of a firm (Bagchi 1997).

Vendor managed inventory (VRP)

This is an alternative to a traditional buyer/supplier relationship. VMI is a concept that has been around since the 1980s, but is only now, in conjunction with SCM, becoming a method for increasing the efficiency of inventory control in a supply chain.

Applying a VMI concept presupposes that a supplier has access to information regarding its customer's production plans, inventory, ordering and sales forecasts. After this stage of openness is achieved, it becomes the responsibility of the supplier to manage and replenish the customer's inventory. The parties agree on more precise guidelines for minimum and maximum inventory levels, as well as for when ownership is transferred and payments are made.

The information exchange between parties can occur through the traditional EDI transfer of information, or be facilitated by exchanging XML files over the Internet. The advantage for the supplier is that it can organize its production planning and materials supply far in advance and thereby utilize its capacity more efficiently and with a more stable workload. For the customer firm, the primary benefit is that the administration of purchasing can be relieved of a number of routine acts, goods receiving control, as well as a reduction in the risk of running out of inventory. The work hours thus released can instead be deployed to the development of strategic alliances and the purchasing of critical goods.

In comparison to the traditional customer/supplier form of cooperation, VMI offers the opportunity of eliminating duplicate functions between vendors and customers, delays in deliveries, out-of-data merchandise and so on. At the same time, VMI gives suppliers the chance to optimize their internal capacity utilization and material flow.

Quick response

This refers to the close coordination between a manufacturer and a retailer for the deployment and control of retail inventories. It consists of: 1) the retailer capturing and sharing accurate and timely sales and inventory data with the manufacturer (usually via POS scanners and EDI); and 2) the manufacturer having the authorization to initiate automatic stock replenishment in line with pre-established target inventory levels.

Efficient customer response (ECR)

This is a supply chain process that typically links a firm's inventory replenishment directly to what customers want. ECR is widely popular in the retail industry. The idea of efficient customer response is to deploy electronic transaction processes through an entire supply chain, that is, EDI, category management, continuous replenishment, cross-docking of goods and supply chain partnerships.

ECR is an innovative strategy developed by the retail industry for streamlining the retail supply chain. It is a strategy in which the retailer, distributor and supplier trading partners work closely together to eliminate excess costs from the retail supply chain with a focus particularly on four major opportunities to improve efficiency: 1) optimizing store assortments and space allocations to increase category sales per square foot and inventory turnover; 2) streamlining the distribution of goods from the point of manufacture to the retail shelf; 3) reducing the cost of trade and consumer promotion; and 4) reducing the cost of developing and introducing new products.

ECR focuses on cooperation between the retail link and retail vendors. A closer cooperation between these parties should create the possibility of direct customer response throughout the supply chain. The goal is to increase efficiency in the flow of goods and to utilize the data the retail stores' electronic cash registers collect with a focus on four specific core areas:

1. efficient store assortment (category management);
2. efficient replenishment;
3. efficient promotion (sales campaigns);
4. efficient product introduction.

The core area for which the greatest potential for savings is identified will be connected with optimizing the delivery to retailers, that is, inventory management and replenishment. These are also the areas that have later been focused on during the application and implementation of the ECR concept. ECR builds upon existing concepts such as VMI and Quick Response (QR).

At the centre of ECR is the sharing of sales information with suppliers, something that EDI has begun. But ECR is much more than this. It involves JIT deliveries and partnership sourcing. Instead of sending orders to suppliers as stocks get low, retailers have the task of generating orders to the manufacturers. Both parties jointly agree on the forecasting techniques to use, and the supplier takes over the responsibility for replenishment using up-to-date sales information from the retailer. It also involves the retailer informing the supplier about sales promotions that will affect the volumes sold and thus those ordered. These systems, which are based on trust and confidentiality, may not work in all circumstances. Both suppliers and retailers must gain from the systems and it may be that a critical mass needs to be reached for benefits to be shared around.

Postponement

This is about delaying the time of shipment and the location of final product processing in the distribution of a product until a customer order is received (Ballou 2004). The idea is to abandon the practice of anticipating 'when' the demand will occur and 'what' will be demanded.

There are two important principles when practicing JIT in inventory management. The first concerns working on the basis of customers' orders rather than starting with forecasts. This aspect is referred to as the postponement principle. The second focuses on the sharing of information in the logistics processes, which is referred to as the bullwhip effect in a supply chain. The future development points in the direction of mass customization. This means that production is customized, the production facilities are set up for greater efficiency, and final assembly can easily change from one product line to another.

This form of production demands very close interaction between the parties along an entire supply chain. As production is controlled by customers' orders, it is difficult for firms to have finished inventory waiting. On the other hand, firms can prepare for the production of individual customer orders by building a stock of components and standard goods, which can all enter the final assembly process, when actual customer orders are received.

Customized production is based on two main principles that are closely connected. The first is modularization and the second is postponement. Modularization refers to a finished product being prepared and individualized from different components with a well-defined interface. By combining the components to satisfy customers' needs for specific products, it is possible to manufacture a large number of different products with a limited number of components. A customer can, for example, order a PC comprising parts selected from a number of choices of hard disks, processors, control systems, software suites, screens and a variety of accessories. The finished customer orders appear as individualized orders with unique capabilities, but are in reality manufactured from a number of standard components (Li et al. 2007, 2008).

Collaborative planning, forecasting and replenishment (CPFR)

This is a collaboration process whereby supply chain trading partners can jointly plan key supply chain activities from the production and delivery of raw materials to the production and delivery of final products to end customers. CPFR encompasses business planning, sales forecasting and all the operations required to replenish raw materials and finished goods. The objective of CPFR is to optimize a supply chain by improving demand forecast accuracy, delivering the right product at the right time to the right location, reducing inventories across the supply chain, avoiding stock-outs and improving customer service. Basically, this can be achieved only if the trading partners are working closely together and willing to share information and risk through a common set of processes.

The CPFR process is divided into the three activities of planning, forecasting and replenishment. Within each activity there are several steps:

Collaborative planning:

- negotiates a front-end agreement that defines the responsibilities of the firms that will collaborate with each other;
- builds a joint business plan that shows how the firms will work together to meet demand.

Collaborative forecasting:

- creates sales forecasts for all the collaborating firms;
- identifies any exceptions or differences between firms;
- resolves the exceptions to provide a common sales forecast.

Collaborative replenishment:

- creates order forecasts for all the collaborating firms;
- identifies exceptions between firms;
- resolves the exceptions to provide an efficient production and delivery schedule;
- generates actual orders to meet customer demand.

The value of CPFR comes from an exchange of forecasting information rather than from more sophisticated forecasting algorithms to improve forecasting accuracy. The fact is that forecasts developed solely by a firm tend to be accurate. When both the buyer and seller collaborate to develop a single forecast, incorporating knowledge of base sales, promotions, store openings or closings, and new product introductions, it is possible to synchronize buyer needs with supplier production plans, thus ensuring efficient replenishment. The jointly managed forecasts can be adjusted in the event that demand or promotions have changed, thus avoiding costly corrections after the fact.

Accurate information about inventory levels, orders, production and delivery status will make the management of logistics activities more effective. In particular, abundant information:

- helps reduce variability in a supply chain;
- helps suppliers make better forecasts, accounting for promotions and market changes;
- enables the coordination of manufacturing and distribution systems and strategies;
- enables retailers to better serve their customers by offering tools for locating desired items;
- enables retailers to react and adapt to supply problems more rapidly;
- enables lead time reduction.

ENABLING TECHNOLOGY

In JIT logistics, it is necessary for firms to keep track of the status of production, procurement of materials and delivery of goods to meet customer requirements. To this end, firms must have information on: where the goods are located, what transport and warehousing is in use, what stock to keep and when to reorder. Also, if it all goes wrong, what resources are available to mitigate problems and costs? There are several enabling technologies that can aid firms to resolve these issues.

POS system

A product of the auto identification technology, this is mainly used in retail outlets to provide accurate inventory control. A POS system enables precise tracking of Stock Keeping Units (SKU) being sold and facilitates inventory replenishment (Bowersox et al. 2002). It also helps to eliminate the bullwhip effect associated with inventories and significantly reduces costs (Coyle et al. 2003).

Enterprise resources planning (ERP)

This is an information system that gathers data from across multiple functions in a firm. ERP helps to monitor orders, production schedules, raw material purchases and finished good inventory. Such an ERP system supports a process-oriented view of business that cuts across different functional departments. For instance, an ERP system can view the entire order fulfilment process and track an order from the procurement of materials to fill orders to the delivery of finished products to customers.

ERP systems provide integrated operations and reporting capabilities to initiate, monitor and track critical activities such as order fulfillment and replenishment, accounting transactions and capacity requirements. Financial, accounting and human resources features are also available so that firms can manage all the in-house activities with ease. Continuous ERP development is aimed at providing additional support functions such as CRM, collaborative planning, forecasting and web-enabled applications for firms as well (Bowersox et al. 2002). ERP systems come in modules that can be installed on their own or in combination with other modules. There are usually modules for finance, procurement, manufacturing, order fulfilment, human resources and logistics. The focus of these modules is primarily on carrying out and monitoring daily transactions. ERP systems often lack the analytical capabilities needed to optimize the efficiency of these transactions.

Advanced planning and scheduling (APS)

This is a set of integrated modules that advance the capabilities of ERP to the next level by applying artificial intelligence to plan for firms' operations. Advanced Planning and Scheduling (APS) consists of four major components, namely demand management, resource management, requirements optimization and resource allocation. ERP forms the basis of APS (Figure 4.6). APS systems are highly analytical applications whose purpose is to assess plant capacity, material availability and customer demand. These systems then produce schedules for what to make in which plant and at what time. APS systems base their calculations on the input of transportation level data that are extracted from ERP or legacy transaction processing systems. They then use linear programming techniques and other sophisticated optimization algorithms to create their recommended schedules.

Demand planning system

The focus is on coordinating sales forecasts from all the players at the POS through production to raw material suppliers, so that everyone is working to the same plan, rather than creating unnecessary safety stocks. These systems use special techniques and algorithms to help firms forecast their demand. These systems take historical sales data and information about planned promotions and other events that can affect customer demand, such as seasonality and market trends. APS uses these data to create models that help predict future sales. Another feature that is often associated with a demand planning system is revenue management. This feature lets a firm experiment with different price mixes for its different products in the light of the predicted demand. The idea is to find a mix of products and prices that maximizes total revenues to the firm. Based on the data gathered by an ERP system, a demand planning system develops requirement projections by generating sales forecasts according to sales history, scheduled orders, marketing activities and customer information, and passes the forecasts to resource management. Resource management gathers and records the resources and constraints in a supply chain system. With demand forecasts, resources and constraints in place, resource optimization is ready to compute the best use of resources for firms. Finally, resource allocation refines the suggested resource assignments and transfers the decisions to the ERP system, which will

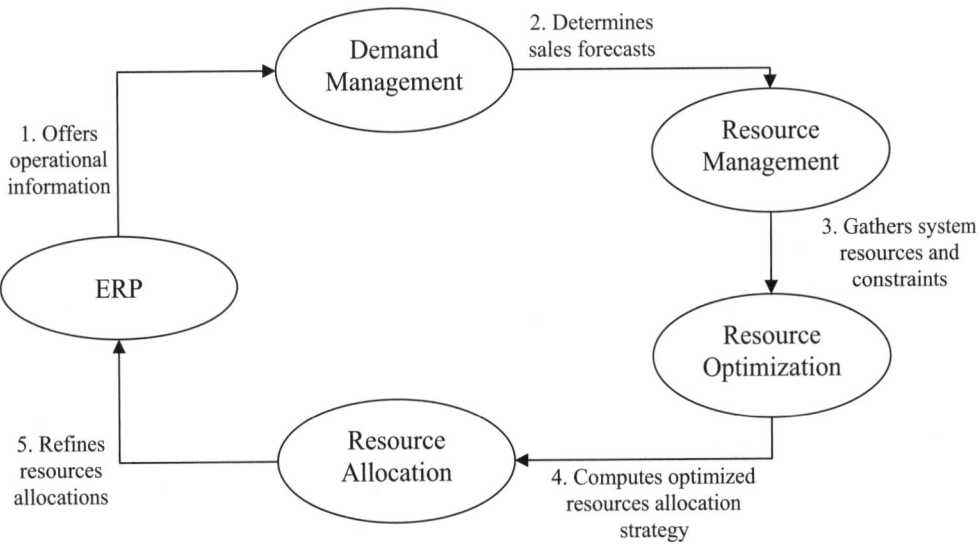

Figure 4.6 Advanced planning and scheduling

record the planned allocation activities. Firms in the travel industry, for example, airlines, rental car agencies and hotels, are major users of revenue management techniques. Other industries are catching up on the value of revenue management using demand planning system.

Transportation Management

When products are completed, they are delivered to customers through different modes of transportation. Delivering products to customers at the right time, quality and quantity requires a high level of planning and cooperation between a firm, its customers and the various distribution elements or services employed (for example, transportation, warehousing and break-bulk or repackaging services).

Transportation is what allows products to move from a point of origin to a point of consumption throughout a supply chain, and is thus responsible for creating time utility and place utility. Time utility is created when customers get products delivered at precisely the right time, not earlier and not later. The transportation function creates time utility by determining how fast products are delivered and how long they are held in storage prior to delivery. Place utility is created when customers get products delivered to the desired location, and again, it is the transportation function that accomplishes this. Thus, transportation in a supply chain is extremely important in that products must be routinely delivered to each supply chain partner at the right time and to the right location. When mistakes occur in deliveries along the supply chain, more safety stocks must be held and customer service levels deteriorate, eventually causing higher costs and lower customer satisfaction for end consumers.

Transportation is necessary for both moving purchased goods from suppliers to buying organizations and moving finished goods to customers. An efficient and effective transportation system is needed for commerce to occur in any industrialized society (Lun et al 2009). Products have little value to customers until they are moved to customers for consumption. For supply chains in particular, transportation is what creates the efficient flow of goods between supply chain partners, allowing profits and competitive advantages to be maximized.

Transportation management decisions typically involve a trade-off between cost and delivery timing or customer service. Motor carriers (trucks) are typically more expensive than rail carriers but offer more flexibility and speed, particularly for short routes. Air carriers are yet more expensive but much faster than any other transportation mode. Water carriers are the slowest but are also the least expensive. Finally, pipeline transportation is used to transport oil, water, natural gas and coal slurry. Many transportation services offer various modal combinations, as well as warehousing and customs-clearing services. In JIT logistics, transportation management is critical to the overall success of the logistics processes.

Transportation management tends to be viewed by manufacturers as 'putting products on a truck, train, plane or bus'. The requirements of JIT logistics for transportation management include every step to move materials from the hand of the last value adder at a supplier location to the hand of the first value adder at a customer location. The desired outcome of transportation is customer service. In order to provide the desired level of customer service, firms must identify customer requirements and then provide the right combination of transportation, storage, packaging and information services to successfully satisfy those requirements. Through frequent contact with customers, firms develop CRM strategies regarding how to meet delivery due dates, how to successfully resolve customer complaints, how to communicate with customers and how to determine the distribution services required. From a SCM perspective, these customer activities take on added importance, since second-tier, third-tier and final product customers are ultimately dependent on the distribution outcomes at each stage within the supply chain.

THE NATURE OF TRANSPORTATION

A firm's transportation systems comprise of the physical links connecting geographically dispersed organizational actors, including customers, suppliers of raw materials, plants, warehouses and channel members. Transportation often represents a significant expenditure for firms – it often amounts to two-thirds of the total logistics costs – therefore it is vital for firms to ensure that their transportation systems are in line with the rest of the activities in the logistic mix to achieve efficiency as well as effectiveness. For instance, an efficient transportation system will add value to a firm by creating time and place utilities, that is, making products available at the designated time, thereby granting the firm a competitive advantage. An efficient transportation system also facilitates accurate production scheduling since it helps to feedback market demand.

TRANSPORTATION MODES SELECTION

The basic modes of transportation available to firms include motor, rail, air, water and pipeline. Each mode has different economic and technical structures and provides different qualities of transport service. Besides single mode transport, intermodal transport is common. Typical combinations include, for instance, rail-motor, motor-water, motor-air and rail-water. Intermodal transportation is sometimes preferred because it provides tailor-made or lower cost services that a single transport mode may not be able to provide.

In order to develop a sound transportation plan, a firm must first consider the characteristics of its products to be transferred. A physical product possesses tangible attributes that have an impact on transportation management. These attributes are:

- weight;
- volume;
- shape;
- perishability;
- flammability;
- substitutability;
- breakability.

The attributes above make a direct impact on a firm's transportation strategy. Combining the attributes creates meaningful indicators for not only transportation, but also warehousing, inventories, materials handling, packaging and order processing. Table 4.7 shows possible combinations.

Table 4.7   Product characteristics and their implications for transportation

Combination	Attributes used	Implications
Weight-bulk ratio	Weight, volume, shape	Increase in weight-bulk ratio implies lower transportation and storage costs – for example, rolled steel, wood
Value-weight ratio	Value, weight	High value-weight ratio leads to higher storage costs and relatively low transportation costs – for example, diamonds
Substitutability	Substitutability (that is, whether the consumers perceive significant differences between the firm's product and that of others)	High substitutability implies an efficient or reliable transportation system or high storage costs to ensure the products are available for sale – for example, soft drinks, junk food
Risk characteristics	Perishability, breakability, flammability	High risk generally requires higher transportation and storage costs to protect the product from damages – for example, fresh fruits, electronic devices

While transportation modes play a crucial role in the management of logistics nodes and the logistics mix, carriers must be able to provide links among logistics nodes and match the rest of the logistics activities. A carrier's transportation costs consist of:

- the rates – the price charged;
- required minimum weights – the minimum charge;
- loading and unloading facilities – for example, at wharf and terminals;
- packaging and blocking – for example, for fragile products such as glass;
- damage in transit – for example, refrigeration, icing and ventilation for perishable goods such as fresh fruit;
- other special services provided by the carrier – for example, stopping in transit.

Transportation rates often become one of the first considerations for firms as small shipment sizes may lead to higher transportation rates, given the volume-based pricing structures (Bagchi et al. 1987). Transportation selection is not merely an evaluation of cost, but a careful consideration of:

- Transit time and reliability – transit time is the total time required to transfer products from the point at which the sender makes the goods available for dispatch to the moment that the goods reach the hand of the recipient. Reliability is the consistency of the transit time that a carrier provides.
- Inventory and stock-out costs – inventory and stock-out costs are directly affected by transit time. Longer or unstable transit time suggests higher inventory levels.

- Product differentiation – if a carrier secures reliable transit times for a firm, the firm will be able to distribute its products to the intermediaries according to predictable transit times. The reliability that the firm provides for its customers eventually becomes the firm's competitive advantage because its customers can depend upon its services to maintain lower inventories.
- Capability and accessibility – capability is a carrier's ability to provide the equipment and facilities, for example, equipment with controlled temperatures or humidity, that moving a particular commodity requires. Accessibility refers to a carrier's physical access to sites. Examples of accessibility factors of a carrier include geographic limits of a carrier's route network and the operating scope of the regulatory agencies.
- Security – security is the ability of a carrier to preserve the products in the same condition as they were when moved to the carrier. Unsafe transportation service may lead to loss of sale or lower productivity.

WASTES IN TRANSPORTATION

JIT requires changes in transportation practices. Under JIT, logistics systems must deliver products on time so as not to disrupt operations (Vonderembse et al. 1995). On-time delivery is critical to support JIT initiatives and reduce inventory (Gentry 1996). Under JIT, delivery schedules are determined on the basis of the customer's schedule, as opposed to the supplier's schedule (Ansari and Modarress 1990). It requires responsive and speedy transportation to meet the requirements. Designing and building a distribution network is one method of ensuring successful product delivery. Again, there is typically a trade-off between the cost of the distribution system's design and customer service. For example, a firm may utilize a large number of regional or local warehouses in order to deliver products quickly to customers. The transportation cost from factory to warehouse, the inventory holding cost and the cost to build and operate warehouses would be quite high, but the pay-off would be better customer service flexibility. On the other hand, a firm may choose to operate only a few highly dispersed warehouses, saving money on the inbound transportation cost, the inventory cost and the warehouse construction and operating cost but then having to be content with limited customer service capabilities. Customer desires and competition levels play important roles in this network design decision.

Anything that disrupts the transportation processes are considered wasteful in JIT logistics since they increase the transportation time. Unutilized resources should also be eliminated since the transportation system should always be refined to make the best use of resources, for example, trucks and cargos, so that items are transferred in a reliable and efficient manner. The list of possible wastes in transportation can therefore be long. Items in the following list are self-explanatory and they serve to offer a quick overview of the potential wastes in transportation management.

- idle time in waiting for deliveries;
- damaged products due to inadequate packaging;
- unutilized vehicles;
- excessive finished goods inventory;
- goods are unavailable for consumers.

TRANSPORTATION SERVICES AND LOGISTICS SERVICE PROVIDERS

Table 4.8 Transportation selection criteria and the implications for firms and service providers

Criterion	Implications for firms as shippers	Implications for service providers
Reliability	On-time pickup and delivery	On-time loading and unloading
Lead time	Total time in transit	Reduced idle time in transit
Cost	Freight rates and costs	Cost to provide required services
Frequency of service	Frequent pick-up and delivery	Fright volume
Damage and loss	Reduced damage and loss	Use of proper packaging and handling equipment
Location coverage	Service locations	Ability to reduce backhauls due to customer locations
Flexible pricing	Rooms for rate negotiation	Willingness to provide value-for-money services
Financial stability	Financial position of service providers	Financial position of customers
Claim service/processing	Prompt settlement of claims and complaints	Responsive payment for services rendered
Shipment tracing	Ability to trace cargoes	IT-enabled operations
Reputation	Carrier reputation for dependable services	Shipper reputation for having the cargoes ready for logistics services
Special handling	Special requirements for services and equipments	Capability to provide different types of logistics services

TRANSPORTATION INFORMATION AND SERVICES

In the JIT context, transportation includes every step required to move materials from the hand of the last value adder at a supplier location to the hand of the first value adder at a customer location. Similar to the case of suppliers, firms are encouraged to establish long-term partnerships with their carriers so that:

- mutual efforts can be sought to resolve problems;
- reliable pickup and deliveries are promoted;
- carriers are encouraged to invest in service improvement with constant and relatively large transaction volumes.

While establishing long-term cooperating relationships with carriers is the first step in ensuring smooth transportation services, in order for JIT to be successful in transportation logistics, firms will also need to provide their carriers with the following information as shown in Table 4.9 (Clouse and Gupta 1990).

The use of Internet services can reduce the cost of transactions between shippers and carriers. Instead of spending time on inter-organizational communication with the carriers by telephone and fax, shippers can manage the simpler aspects of transactions through the carrier's website (Lu et al. 2006). In addition, Internet services can help reduce the cost and time of putting together and handling shipping documents. Putting together commercial invoices, bills of lading, manifests and customer documents is an enormous expense for exporters, importers and shipping lines.

TRANSPORTATION SECURITY

Transportation security is increasingly becoming a concern to both the public and private sectors, particularly after the disastrous 9/11 event at New York City on 11 September 2001 (Sheffi 2005). The potential theft and mass destruction embedded in the shipments through the transportation process will disrupt JIT logistics, causing wastes and service deterioration to different parties in the supply chain. The resulting wastes can be huge. Such wastes may include inventory costs due to obsolescence, markdowns, stock-outs and penalties for non-delivery of goods. It is necessary for firms to prepare for transportation security and respond with proposals to create confidence in logistics security, while maintaining smooth flows of goods and services in a global supply chain (Lee and Whang 2005). The events and aftermaths of 9/11 have not only called for heightening of global transport security, but also triggered a fundamental shift in the way transport policies and regulations are being drafted, managed and implemented. Although the international transport community has long reacted to the influence of commercial pressures, whereby security is only considered during times of huge claims and crippling insurance premiums, the growing pressures of regulatory

Table 4.9 Information desired for better transportation services

Information	Purpose
Previous freight volumes	To determine equipment dedication and rate structures
Transit time requirements and operating procedures	To be well aware of shipment requirements
Detailed service needs such as break-bulk operations, consolidations and warehousing	To determine equipment utilization, routing and transit times
Traffic flow patterns	To determine schedules for shipments
Nature of freight	To schedule the proper type of equipment and make decisions about its loadability
Backhauls	To avoid non-productive empty miles for the trucker

bodies (for example, customs) have prompted transport industry organizations to heighten security measures in their operations.

Disruptions in the transportation process can take place in numerous ways and affect companies in an unanticipated manner, and at any time. The closure of the US air space after the terrorist event in 2001, the port labours' strike in California in 2002 and the outbreak of SARS in 2003 are examples of unexpected incidents that have disrupted the transportation process in the global supply chain. Assuring transportation security involves necessarily ensuring the security of trade between countries and trading partners around the world. This involves local governments, export and import firms, transport logistics firms, manufacturers, as well as foreign governments and trading firms around the world (Lai and Cheng 2004). Security assurance across the global supply chain has increasingly become a critical success factor for many firms in the evolution of international trade (Lai 2004). Following the 9/11 tragedy, a number of initiatives are underway focusing on security in ports and ships (International Ship & Port Security Code – ISPS), customs inspections in international ports (Container Security Initiative – CSI), and international cargo flows in the global supply chain (Customs & Trade Partnership against Terrorism – C-TPAT). The International Maritime Organization (IMO) and other groups such as the World Customs Organization (WCO) have jointly supported processes that enhance regulatory coverage of safety and security within the world trading system. As a result of the efforts of the IMO, a number of security measures have been formalized, including changes of the Safety of Lift at Sea (SOLAS) convention that specifically addresses ship security with updated requirements for compliance with the ISPS Code.

The security issues for transportation emerge from a number of factors, namely:

- Cargo – using cargo containers to smuggle people and/or weapon (of a conventional, nuclear, chemical or biological nature).
- Carrier – using aircraft/vessels as weapons or means to launch an attack (including hijacking an aircraft or sinking a vessel to disrupt infrastructure).
- People – using fraudulent cabin crew/seafarer identity to support terrorist activities.

Containers move along a network of transportation nodes and links. The nodes are physical locations, such as container terminals and depots, where containers are handled and transferred from one transport mode to another. The links between nodes are served by modes of transport such as roads, rails and water, and supported by infrastructure such as highways, rail tracks and ports. The risk of a security breach at any one of the nodes or links can compromise the security of the entire contain transport chain. For instance, the stuffing area of a container transport chain is crucial to container transport security because it represents the last point in the container transport chain where the physical contents of a container can be visually identified and reconciled with related documents such as container load plans. Once the container is sealed, all the information on the contents of the containers cannot be verified until the container is reopened by a consignee. As such, shippers play a critical role in container transport security by generating accurate and complete sets of data about the cargoes stuffed in the containers. Moreover, containers moving through different modes of transport face many security challenges as the transport carriages involve multiple stops (for example, container yards and terminals) where containers are stored and handled, and open transport infrastructure, which can be accessed by different organizations and individuals (Lun et al. 2008).

While it would be practically impossible to inspect every container from every aircraft/vessel that arrives at the destination port, it is possible to implement certain practices across the supply chain that can help minimize such risks as terrorism, theft and smuggling, whether conventional or non-conventional, reaching our shores. By doing so, it will contribute to not only the security of the cargoes, but also the safety of the people and facilities handling the logistical flows of goods in the supply chain (Bichou et al. 2007). In essence, the idea is to make supply chain entities and governments around the world liable to ensuring the security of the goods and modes of transportation while in their custody and the safety of the people involved in the processes. It is much cheaper and less risky to assure security (just like taking preventive measures to assure quality) if security processes are implemented throughout the supply chain, instead of security being inspected at the end of the process, that is, at the ports.

In view of the importance of ensuring transportation security, Russell and Saldanha (2003) proposed five tenets of security-aware logistics and supply chain operations. These five tenets are summarized below:

- Tenet 1: firms need to partner with local and foreign government organizations that impact the movement of freight.
- Tenet 2: now more than ever, firms need to know their overseas trading partners and take responsibility for securing their cross-border supply chains.
- Tenet 3: firms need a mode-shifting capability to accommodate unexpected delays, interruptions and disasters.
- Tenet 4: firms need to develop a suite of communication channels and media to manage crises.
- Tenet 5: there is a need to adopt the military concepts of agility, reservists and pre-positioning for the management of business logistics and the supply chain in the new environment.

As firms continue to build capabilities to react appropriately in a crisis and keep terrorist threats to a minimum, changes to long-term SCM strategies are inevitable. Firms clearly need to adopt new tenets of operations. Improving management of relationships with local and foreign governments will facilitate awareness of and involvement in government decision making impacting private sector logistics and SCM. It is also highly desirable for logistics and supply chain professionals to pay more attention on customer and supplier relationship management. Mode shifting will play a crucial role in accommodating unexpected delays, interruptions and disasters. Moreover, in supporting these additional efforts and changing practices, firms need to consider a suite of communication channels and media to manage crises. Firms also need to embrace contingency planning and its linkages to the military roots of logistics to ensure preparedness. Finally, firms need to be aware of, and contribute to, government's work in securing the country against future terrorist attacks, both fulfilling their duty as corporate citizens, and ensuring supply chain security and business continuity for themselves.

TRANSPORTATION MANAGEMENT TOOLS

There are several management tools that are useful for firms to design their transportation for JIT logistics. These management tools include:

Network design

Developing a transportation network is concerned with the arrangement of various means of transportation, for example, the number of levels in the distribution network, the number of distributing facilities, deployment of inventory in the network and location of logistics facilities. The goal of performing network design is to optimize the inventory carrying costs and the transportation costs while aiming to deliver the items on time. Like other logistics processes, there are trade-offs among the cost factors in a transportation network. For instance, more distribution nodes may lead to higher inventory costs (as more inventories will be stored in the distribution outlet) but lower transportation costs (since items can be transported from a choice of outlets). In order to develop an optimized network design, firms should first of all develop a multi-facet database that includes all the transportation-related details. With respect to all the nodes in the network, details like travelling time, inventory carrying and transportation costs, response time requirement, product availability and product demand, and throughput and storage constraints need to be stored in the database. With the information on hand, firms will be able to configure network design alternatives and derive an optimal plan with the help of mathematical programmes and decision tools (Frazelle 2002).

Scheduling and routing

Routing is about minimizing total route costs, number of routes, total distance and time, while scheduling focuses particularly on 'time' constraints. Routing is often depicted as one of the most challenging computational problems and heuristics are employed to support most of the routing software. A good routing solution should include backhauling, that is, interleaving pickups and deposits on a single route to maximize vehicle utilization. More sophisticated routing solutions like continuous moves, which literally maintain each vehicle in continuous operations, between nodes can also be derived. In addition to deriving a routing plan, scheduling will refine the plan by actions like avoiding peak hub times (Frazelle 2002).

Quality audit

Claim rates, customer services, billing-error rates, safety record, transit times, equipment availability, dispatching operations and truckers' annual quality improvement programmes form a valid set of quality criteria for firms to evaluate the performance of their carriers (Clouse and Gupta 1990).

Transportation management system

This is concerned with determining what methods of transport at what cost are available to shippers, what costs and timings are associated with each route, the order in which to load transport, the optimization of multi-segment deliveries and compliance with documentation required for customs and shippers. This system enables firms to compare different modes of transportation, routes and carriers, and use the comparisons to create transport plans for the items to be delivered. The software for such systems are provided by system vendors, and the content vendors provide the data required, for example, mileage, fuel costs and shipping tariffs, for operating the system.

Warehouse management system

This is a system used to optimize the storage and picking of goods in a warehouse, including compliance with carrier and customer documentation standards. The system supports daily warehouse operations by keeping track of inventory levels and stocking locations within a warehouse, as well as the actions needed to pick, pack and ship product to fill customer orders.

ENABLING TECHNOLOGY

In recent years a host of technological innovations and enabling technologies for transportation management have emerged. If properly deployed, these enabling technologies can help firms to attain the dual goals of cost reduction and service improvement in transportation for JIT logistics. Examples of these technologies are provided below:

Electronic Product Code (EPC)

It is essential for firms to possess the capability to track and trace their product items in the transportation processes. To this end, product information processing is probably the most salient issue concerned with automatic processing for product track and trace. Electronic Product Code (EPC) is a type of auto identification system, which requires the placement of computer readable codes on item surfaces, cartons, containers and even rail cars to facilitate the collection of production information. EPC digits are represented as vertical lines with different line widths. Readers will be used to scan the mounted bar code in order to convert it to the respective product code.

Integration of the internal functions of an organization is important and this should precede the forging of external links with suppliers and customers. A high level of integration is required in supply chains, in which information sharing amongst intra- and inter-organization processes facilitates the autonomy of operations in the supply chain. The bullwhip effect in the supply chain also illustrates the importance of sharing information in supply chains. As a popular standard for identifying and tracing products, EPC is useful for managing supply chain processes, reducing errors in data input, enhancing communication between firms and increasing the traceability of trade items in supply chains, all of which are fundamental to effective SCM in any industry. To illustrate, an EPC tag is an identifier, in the form of a label, licence plate or computer chip, which can be identified by a reader that is connected to a system for information about the tagged object. can be deployed across the functional processes of an organization – from sourcing to customer service, and across a range of tasks – in support of internal operations, as well as facilitating the coordination of the supply chain. Potential uses of EPC for business logistics include: 1) asset management, tracking and maintenance, 2) volume planning and automated data capture through shipping route, 3) shipment route tracing and identification of package content, and 4) automated customs.

The purpose of using EPC is to facilitate flows of physical products in the supply chain. The ability of firms to identify specific trade items accurately and in a cost-effective manner is critical to managing supply chain processes at the operational level. To ensure the accuracy and integrity of information, the best method is to minimize human intervention in the processes of capturing and transferring data. A popular method of entering data, EPC in the form of computer-readable bar codes (or in other forms, such as Radio frequency identification (RFID) tags) enables different parties in the supply chain to identify products, organizations, locations and shipments of goods with a unique number represented by a bar code. The information contained in a bar code can be retrieved electronically by a computer and transmitted to other computers via communications systems, for example, EDI. Unlike other information technologies that often require manual input of data, the operational nature of EPC requires no manual input of data, thus preserving the integrity of the data and incurring a lower operations cost. Recently, the need for heightening the security of ocean containers after the disastrous event of 9/11 has called for the adoption of EPC, for example, RFID (which is elaborated on below), to prevent any container from being used as a weapon of destruction without impeding the smooth flows of containers. Recognizing the importance of port security, Hutchison Port Holdings Group, which is one of the world's largest independently owned port investors, developers and operators, has installed RFID sensors in four of the group's major ports, to test the feasibility of using RFID to improve the security of its container ports. Moreover, fast and accurate flows of information resulting from the adoption of EPC enable firms in the supply chain to streamline their processes and enhance their ability to handle physical product flows in a cost-effective manner. A study of Consumer Packaged Goods (CPG) companies, for example, The Home Depot, Procter and Gamble, Unilever and Wal-Mart, found that CPG companies embracing EPC to collaborate with their trading partners can generate performance gains in growth, profitability and consumer satisfaction (Chappell et al. 2002). In such cases, EPC contributes to the fast and accurate capture and transfer of data, leading to increased traceability of trade items in the supply chain, that is, the ability to identify and trace products, which is fundamental to effective SCM in any industry.

Global positioning system (GPS)

The technology that will control the future of transportation is efficient information flow and automated data collection. The key issue in the future will be providing various information on a shipment, including its location, where it needs to go and when it will get there. As cycle times shrink and inventory-lean manufacturing processes become the norm, transportation buyers will need to be focused on where a shipment is in the distribution pipeline, rather than with the physical shipment itself. Global positioning system (GPS) is a synthesis of satellite tracking, computer, communications, wireless and Internet technologies. It provides fleet managers with the ability to determine vehicle positions with high-quality, street-level detailed maps on computer workstations. Details on cargo and vehicle security and route optimization are available. Even driver monitoring and two-way communication becomes possible with GPS. Data are sent via a compact vehicle-mounted unit that picks up GPS signals from orbiting satellites. Extensive reports on idle time, pickup times, speeding, event history and routes taken are generated with ease (Alternburg et al. 1999).

By synthesizing satellite tracking, computer, communications, wireless and Internet technologies, GPS offers fleet managers a powerful tool to increase productivity and achieve major cost savings and operating efficiencies. GPS provides fleet operations with the ability to display the locations of vehicles on their PC screens using high-quality, street-level detailed maps, cargo and vehicles security, route optimization and driver monitoring. GPS providers are utilizing the Internet as a communication medium, offering lowest cost solutions for firms wishing to monitor vehicle fleets around the country from a central location.

RFID

This is an emerging technology that has been increasingly being used in logistics and SCM in recent years. RFID technology can identify, categorize and manage the flow of goods and information throughout a supply chain (Lai, Wong and Cheng 2006). Basically, it is made up of two components, namely a transponder, which is located on the object to be identified, and a reader, which, depending upon the design and the technology used, may be a read or write/read device (Finkenzeller 2003). The development of RFID technology has not only saved time and labour costs by eliminating the need to manually scan the bar codes of goods, but has also solved the problem of goods going out of stock due to confusion in a supply chain and the problem of shrinkage caused by theft. The microchip is capable of storing a vast amount of information and the attached information can be easily converted into compatible formats for follow-up processing. When more information can be recorded and gathered on an itemized level, operational processes can be tracked down to the finest bits of details. Firms should ensure that they are utilizing the data well to develop the most optimal strategy to achieve and sustain the cost and service competitive edges. There are many useful applications of RFID in logistics contexts. For instance, RFID has been applied in container depots, integrated with mobile commerce to: 1) keep track of the locations of stackers and containers; 2) provide greater visibility of the operations data; and 3) improve the control processes (Ngai et al. 2007a). There are also RFID-based traceability systems in the aircraft engineering context to support the tracking and tracing of aircraft repairable items (Ngai et al. 2007b).

World Wide Web (WWW)

The increasing importance and proliferation of the Internet and the World Wide Web (WWW) represents a significant development in the use of Internet services. The Internet is a global collection of computer networks linked together for the purpose of exchanging data and information. It provides an efficient way to search and share information. The WWW has grown to become a collection of many interlinked computers, which function together within the Internet. The Web can handle much information in many different formats such as texts, graphic images and media and it has become the premier means to speed communication between customers and their suppliers, improving service levels and reducing logistics costs (Lai et al. 2008b). There are many transportation-related activities that can be performed via the Web such as cargo tracking, the issuing of electronic bills of lading, the booking of shipping space and access to information on sailing schedules, equipment and tariff rates (Lu et al. 2005). Nowadays, major liner shipping companies such as Maersk, Evergreen, MSC, COSO and OOCL utilize the Web to provide shipping services, which include:

- tracking (cargo tracking, container tracking, email tracking);
- vessel schedules (sailing schedules);
- customs response (export cargo customs checks);
- finding tariff rates (online rate enquires and confirmations);
- booking services (booking space);
- electronic document services (search, view, verify, approve and print B/Ls and waybills, EDI);
- company profilea (history, philosophy, testimonial archive, features and benefits);
- organization (organizational hierarchy, office directory, agencies);
- financial information (financial reports, annual reports, share prices and so on);
- operational reports (safety, environment, operational plan, introduction to the computer system);
- publications (other magazines, papers and reports);
- fleet information (class, capacity in TEUs, width, length);
- container information (types, sizes, capacities);
- special equipment (reefers, flat rack containers, open-top containers and high cubes);
- news (press releases, latest news, news archive);
- market information (trade news, rates of exchanges, world times);
- industry information (shipping research reports or statistics)
- Service routes (service coverage, intermodal);
- local language services (English, Chinese, France, Spanish and so on);
- web support (help, search engine, contact, site map);
- download services (documentation, installation, search tools, help tools, for example, Acrobat reader);
- enquiry services (frequently asked questions, email);
- registration (web registration, data modification);
- web link services (group, industry-related web, government).

Summary

JIT logistics covers many different areas in business logistics. In this chapter we discussed the four major logistics mix elements, namely customer services, order processing, inventory management and transportation, and how firms can attain the goals of JIT logistics for cost reduction and service improvement through better management of logistics activities in these four logistics areas. To this end, we examined the principles for managing these four important logistics mix elements. The possible wastes in customer service, order processing, inventory management and transportation were identified and the ways to reduce waste discussed. The management tools and enabling technologies that are useful for firms to eliminate these wastes and to improve their service performance through the JIT logistics approach were introduced. To achieve cost reduction and service improvement via JIT logistics, it is necessary that firms focus on the quality management issues. The outcome of logistics, that is, the attainment of the 7Rs, is likely to be compromised if the quality issues are neglected. The next chapter will discuss the various quality management issues for the implementation of JIT logistics in firms.

CHAPTER 5
Quality Management Issues

Quality Awareness in Logistics

To implement JIT logistics, quality management issues must not be overlooked. In logistics 'quality' is about the achievement of the 7Rs. It is the responsibility of all the parties involved in the process. Failure to attend to the quality issues will bring a negative impact on the next value-adder, or the customer downstream. To tackle quality issues, quality management systems, particularly Total Quality Management (TQM), have received worldwide attention in the past decade and are being pursued in many nations. Quality management has gained in popularity mainly because of increasing consumer consciousness of quality and growing international competitive pressure. On the other hand, product/service quality has long been regarded as a source of creating cost and service advantages. It has proved to be effective in lowering manufacturing costs and improving productivity (Garvin 1983) and firms across different industries have experienced improved organizational performance, for example, lowered production costs and improved profitability, as a consequence of quality management implementation (Lai and Cheng 2003a). Industry-wide acceptance of quality management has resulted in extensive diffusion of quality concepts.

There are a number of factors that account for the growth of quality awareness in business logistics. First, competitive pressure has prompted many business firms to deliver better value in their product/service offerings due to escalating customer demands for better logistics services to achieve cost reduction and service improvement. As a result, the creation of time and place utilities in the 7Rs, that is, the ability to deliver the right amount of right product at the right place at the right time in the right condition at the right price with the right information, are essential to meet customer requirements in logistics performance. Similar to other business functions, many firms have implemented various quality improvement programmes for their logistics activities with a view to gaining a competitive position for their products/services. Some of them even view quality in the form of a certified quality system, like the ISO 9000 quality management series, as a competitive advantage for managing their logistics activities.

Second, the pressure from the customer groups is characterized by their increased emphasis on speed to market with high quality (Lai et al. 2004). The shortening product life cycles and the proliferating product choices have forced business firms to compete on quality products, consistent product availability and faster product delivery to meet customer demand (Fliedner and Vokurka 1997). Accordingly, business firms are increasingly looking to logistics management to meet the challenges with respect to time- and quality-based competition. Quality management is a means for firms to handle their physical product flows and to ensure that the added value in the 7Rs are delivered to meet the customer expectation in the logistics processes.

Furthermore, quality management has been widely recognized as a potent means for achieving a competitive edge from differentiation across a broad spectrum of business sectors (Lai and Cheng 2003a). Improving product and service quality is increasingly considered as an effective strategy for gaining sustainable competitive advantages (Yeung et al. 2005). Recognizing the potential benefits that improved quality is likely to bring, firms are increasingly embracing quality management principles for the management of their logistics activities.

The Need for Quality in Logistics

To deliver customer value, it is essential for firms to thoroughly understand and effectively manage different aspects of their logistics activities in order to ultimately conform to customer and market requirements in a cost-effective manner. Quality improvement is meaningless without an understanding of the competitive implications of quality. For instance, customers tend to view quality primarily in terms of the level of reliability and the associated logistics costs, whereas firms tend to view quality from a broader perspective embracing such areas as customer service, administration, maintenance, storage and information. A clear conceptualization of quality and a concerted effort to attain superior quality are important for firms to meet the requirements desired by all the parties concerned in performing logistics activities. Divergent interpretations of quality and work process requirements will obscure organizational directions and compromise efforts to improve quality (Garvin 1987). To reach a common quality goal by invoking company-wide efforts for quality improvement, there is an increasing need for firms to have a mechanism, for example, quality management, devised and deployed to support improvement efforts in logistics activities.

Quality management requires firms to develop and implement a corporate-wide culture emphasizing customer focus, CI, employee empowerment and data driven decision making. It is a holistic management approach that strives for CI in all the functions of an organization. Oakland (1993) defined quality management as an approach to improving the effectiveness and flexibility of business as a whole. According to him, quality management is essentially a way of organizing and improving the whole organization, every department, every activity and every single person at every level. The aim is to continuously improve process performance by placing customers at the focal point of operations in order to fully satisfy them. It is a continuous quest for excellence that has to reach every individual within an organization in order to make prevention of defects possible and satisfy customers totally at all times.

In view of the increasing pressure for quality improvements, many firms have started to embrace its principles for logistics management, in hopes of aligning their product/service delivery with customer expectations and improving performance. There is a multiplicity of frameworks in the literature for implementing quality management (for example, Ahire et al. 1996). Nevertheless, the principles of quality management, to a greater or lesser extent, incorporate the following (Sitkin et al. 1994):

- the generating of objective data ('facts') for the systematic improvement of work processes and products as a prerequisite for taking action;
- a focus on key problem areas and customer satisfaction;
- the involvement and empowerment of employees.

Although quality management was first used in manufacturing operations, its widespread adoption has gradually been extended to service operations, including logistics operations. The principles of quality management are applicable and useful for logistics to create customer value at a low cost in many aspects of logistics activities. These include warehouse management, shipment consolidation, logistics information management, order fulfilment and processing, carrier selection and production assembly. Similar to manufacturing operations, quality management can help firms achieve the dual objectives of cost reduction and service improvement in logistics activities by the CI of work processes through procedure design, policy deployment, human resources management and a set of quality management toolkits.

Quality and Quality Costs

A SYSTEMS PERSPECTIVE

A systems perspective to quality is useful to performance improvements in logistics. It provides a holistic view of a firm as a set of interrelated and interdependent parts, each of which via its behaviour and performance interact with and influence the others. In managing logistics activities as a system, one should stop looking at what one part (for example, inventory) or the other (for example, transportation) is doing in isolation, but concentrate on the patterns of interaction and how these patterns affect one another and the overall performance of the logistics activities. A firm can expect to make improvements in the system by evaluating the cost and service trade-offs in the logistics activities and modify their relationships for better system-wide performance.

What is important in the systems view is the management of logistics activities and not the individual activities themselves. This view helps us understand that the success of any one part depends on what lots of other parts do. It should be noted that work does not get done by individuals working as independent agents but by processes that individuals work together to execute. The initial step towards quality improvement in logistics begins with the people who embody the organization. Questions such as the following should be asked by each employee:

- What is my role within the firm?
- How well do I understand the firm's logistics activities?
- Am I able to identify the firm's expectations of the logistics performance?
- What are the current quality requirements?
- How can I improve my work performance to increase the firm's performance and customer satisfaction?

These questions function as a means of identifying how individual performance can be improved to enhance customer satisfaction. The organizational level of quality improvement is made possible through the following steps:

- Identifying customers' expectations is possible through the development of questionnaires, which are sent to established and potential customers of the firm, to survey their opinions in order to create a customer profile. The questionnaire should

focus on determining how well the customer needs are currently being satisfied; seeking customer input concerning how the product, service or logistics-related activities such as delivery or after sales services can be improved upon.
- Assessing organizational capability. Organizational assessment is a prerequisite to establishing overall JIT logistics goals, the capability of employees, equipment and systems to meet the higher JIT logistics standards, the compatibility of organizational culture and beliefs in achieving the new JIT logistics standards. In effect, the assessment should present a snapshot view of the current organizational conditions. Frequently, the assessment will reveal requirements for increased training and purchases of new equipment. Perhaps the most difficult aspect of striving to implement JIT logistics is changing a culture that does not support such an endeavour. An organization's culture is pervasive and practised by all the employees. Firms that focus on achieving quick results and high output may be faced with a formidable task of trying to put JIT logistics into practice. JIT logistics is as much a part of performance as it is an attitude. Acceptance of new ideas and willingness to work towards the successful implementation of JIT logistics requires attitude changes, which must occur before any benefits from JIT logistics will be realized.
- Relating current performance standards with JIT goals. The discrepancy that exists between the current performance levels and the JIT goals that an organization strives to achieve is known as the 'performance gap'. The performance gap exists wherever the performance of one of the elements of the system is unable to meet new performance standards. Examples include failure of the shipping department to deliver a product on time, lack of qualified machine operators, equipment that operates outside specified tolerances or an inventory system that fails to coordinate the flow of WIP inventories.
- Bridging the performance gap. The goal of the organization at this point is to reduce and eventually eliminate the performance gap. The process should commence with the elements identified in the organizational assessment, that is, what can be done to improve performance and processes to meet the new JIT goals? Many of these deficiencies can be eliminated through increasing the level of training, improving communication between employees, for example, through employee involvement programmes and increasing the level of visibility through workplace improvement, and seeking commitment from top managers and all the employees. To ensure the desired behaviour is continued, performance that is compatible with a culture supportive of improvement should be recognized.
- Implementing the performance improvement system. Table 5.1 presents an example of a performance improvement system applying to people, plants and systems. Resistance to change, as discussed in the first chapter, is impossible to avoid; however, it is not impossible to overcome. While top management is responsible for initiating the organizational change, all the employees should become an immediate part of the process in order to realize success. Employees should be involved in identifying the discrepancies that constitute the performance gap. Attitude changes should stem from the top: a difficult lesson learned by many JIT enthusiasts. The failure of many corporate leaders to demonstrate commitment to JIT improvement and support of employees has resulted in abandoning the effort.

Table 5.1 Quality improvement system summary

Identification	Identifying customer expectations	Develop questionnaire to ask customers 'how can the product/service and logistics activities be improved?' Examine strengths and weaknesses of people, plants and systems
Analysis and measurement	Organizational assessment	Determine the performance gap factors Fundamental performance gap factors include the following: • Lack of total commitment • Poor organizing and coordinating effort • Ineffective communication • Disregard of customer expectations • Lack of employee motivation • No employee ownership • Poor relationship with suppliers • Poor housekeeping habits • Lack of worker visibility • Ineffective arrangement of equipment • Defective equipment • Insufficient training programmes
Solution	Bridging the gap	Generate solutions to problems, subject to organizational constraints
Execution	Implementation of solutions	Formulate the performance improvement plan Establish well-defined goals Develop operational goals Assign roles to key teams and individuals Allocate organizational resources Develop contingency plan
Follow-up	Monitoring and measuring performance	Relate performance to operational goals, compare on the basis of: • Timeliness • Availability of resources • Additional problems encountered? • Resort to contingency plans • Need to establish additional plans? • Seek employee opinion and involvement continuously

THE COST OF QUALITY

To improve quality for JIT logistics, the cost of quality should not be overlooked. The cost of quality can be divided into two components, which at first seem to be at odds with one another. The first component includes the cost of producing a product or service that fails to meet established requirements and customer expectations. A firm's failure to supply an acceptable product/service will result in other costs such as the cost of return, lost customers, poor advertising, warranty and service expenses. The second component includes the organizational cost of achieving established performance standards. The cost that is associated with quality is therefore composed of two components:

1. cost of nonconformance – the cost of producing a product or service that fails to meet established requirements and customer expectations, for example, loss of sales, poor reputation, high inventory levels and so on.
2. cost of conformance – the organizational cost of achieving established performance standards, for example, preventive measures, employee training, self-inspection procedures and so on.

The cost of conformance can be further broken down into appraisal and prevention costs, whereas the cost of nonconformance is sometimes regarded as failure costs (Brinker 2000). The appraisal costs include the costs incurred to assess the quality of a firm's products and services. Appraisal costs fluctuate with the nature of each product/service. The prevention costs include the cost of methods employed to reduce or eliminate the costs of appraisal and failure. The failure costs include the expenses incurred as a result of production of less than satisfactory products or services. Common costs that constitute this broad category include scrap, re-work, warranty and litigation expenses. Perhaps one of the most significant costs of failure is the loss of existing and potential customers. Loss of customers can have a severe impact upon a firm in terms of creating a poor reputation that associates the firm with producing shoddy products/services.

The concept of 'value added' plays an important role in reducing the cost of quality. Activities that are commonly carried out but fail to add value to products or services should be eliminated. For example, final product inspection does little to prevent the production of a defective lot of products. Operating within a JIT environment assists in eliminating many of these costs through the establishment of effective methods of operations, employee utilization and other organizational resources. In a JIT environment, poor quality should not be tolerated since all the parties involved are committed to providing quality products and services.

The emphasis will be on prevention rather than detection, thus the cost of supervision and inspection will go down. Prevention will go up because of training and action-oriented efforts. But the real benefits will be gained by a significant reduction in failures, both internal (for example, re-work, return, repair) and external (for example, handling of complaints, servicing costs, loss of goodwill). The total cost of quality will reduce over time.

A GENERAL GUIDELINE TO QUALITY IMPROVEMENT

The approach to quality is a unique endeavour in itself when a firm strives to meet JIT logistics goals. Quality improvement involves organizational changes that require time and resources. The implementation of quality management requires a fundamental change in the way business is conducted; organizations are managed as systems, employees are empowered; focus is placed on customers and a set of effective management techniques (Yeung et al. 2006). To this end, customer needs and expectations are fully focused upon, employees are wholly empowered and each part of the organization is enabled to contribute to CI. The following elements can serve as a general guideline for directing organizational actions towards quality improvement.

- Customer focus – listening to the voice of the customer, staying close to the customer and meeting or exceeding customer desires is basic to quality improvement for JIT logistics. It is important to focus on quality efforts most likely to improve customer satisfaction at a reasonable cost. A customer focus requires a never-ending, intense focus on customers' needs, wants, expectations and requirements and a commitment to satisfying them.
- CI – it is about pursuing small, incremental, manageable improvements of processes continuously and persistently. A commitment is needed for any CI efforts. The focus should be on both the process, as well as the results. One way to gain improvements is to benchmark and adopt best practices. In doing so, firms may need to measure results against anticipated gains and not hesitate to revise strategic plans, programmes or processes accordingly.
- Prevention focus – quality improvement seeks to prevent poor quality in products/services. Prevention starts with a quality programme. Firms that lack process and inventory controls and other fundamental quality culture will certainly fail in any quality improvement quest. A firm must calculate the cost of current quality initiatives, including warranties, waste, re-work, repair, return, problem prevention and monitoring. It should measure these against the returns for delivering quality products or services to customers.
- Employee involvement – if quality improvement for JIT logistics is to be successful, there must be instilled in everyone a strong and profound belief that the responsibility for assuring the JIT goals is shared by everyone in an organization. This implies that any reward or recognition system should be congruent with the quality improvement efforts. Many managers fail to achieve quality improvements for JIT logistics because they have neglected the behavioural aspects. It is essential to promote what behavioural changes are needed and desirable in quality improvement. Quality should be a guiding philosophy shared by everyone in a firm. In such cases, training, team building and other work-life enhancements need to be provided to all the employees. There should also be proper reward and recognition systems that encourage and reward quality improvements for JIT logistics.
- Management commitment – for any quality improvements to succeed, management commitment is a must. Employees must be shown that team and individual efforts are being recognized for performance. The importance of an employee reward and recognition system may not appear significant; however, it is one way to communicate and reward behaviour. Management must fully comprehend the extent and nature of the fundamental changes required to make quality improvement an integral part of a firm.
- Fact-based decision making – measurement based on reliable information, data and analysis is a basis for improvement. It is important to determine what key factors retain customers and what reasons drive them away. Use detailed surveys and benchmarks. Forecast market changes, especially the quality and new offerings by competitors. All these require making decisions based on data, measurement and statistical information rather than opinions.

These guidelines are underpinned by the famous Deming's 14 points for management. These points are:

1. Create consistency of purpose towards improvement of product and service.
2. Adopt the new philosophy.
3. Cease dependence on inspection to achieve quality.
4. End the practice of awarding business on the basis of the price tag.
5. Improve constantly and forever the system of production and service.
6. Institute training on the job.
7. Institute leadership.
8. Drive out of fear.
9. Break down barriers between departments.
10. Eliminate slogans, exhortations and targets for workforce.
11. Eliminate work standards (quotas) on the factory floor.
12. Remove barriers that rob the worker of the right to pride of workmanship. Remove barriers that rob people in management and in engineering of their right to pride in craftsmanship.
13. Institute a programme of education and self-improvement.
14. Put everybody in the company to work to accomplish the transformation.

Deming's theory of management essentially holds that since managers are responsible for creating the systems that make an organization work, they must also be held responsible for the organization's problems. Thus, only management can fix problems, through application of the right tools, resources, encouragement, commitment and cultural change. He viewed that quality was the outcome of an all-encompassing philosophy geared towards personal and organizational growth. He argued further that growth occurs through top management vision, support and value placed on all the employees and suppliers. Value is demonstrated through investments in training, equipment, continuing education, support for funding and fixing problems, and teamwork both within the firm and with suppliers. Use of statistical methods, elimination of inspected-in quality and elimination of cost-based decisions are also required to improve quality.

ROLES OF EMPLOYEES IN QUALITY IMPROVEMENT

Employee participation plays an important role in quality improvement and their total involvement is one of the crucial elements to achieve success in the process. Total employment involvement is about the collective efforts of all the employees directed towards problem-solving and improvement-seeking activities throughout a firm. The underlying principle of Total Employee Involvement (TEI) is that the employees who work closest with customers, products and processes are most knowledgeable in finding ways to improve these three elements. TEI offers a direct and simple means of keeping employees interested in their work through the establishment of employee ownership. Employees are responsible for setting the goals and directions of the operations that they are involved with. They must be encouraged to raise their ideas and concerns and become involved. Successful implementation of TEI requires:

- providing employees with the necessary resources to identify areas of improvement and recommend changes;
- requiring employees to aim for organizational goals rather than individual needs;
- establishing general guidelines and objectives;
- delegating the authority to resolve problems to employees whereas management will be responsible for providing directions.

Quality of Services

Managing the quality of services for logistics has assumed a new level of importance relative to the growing number of firms operating in the service industry. The shift in industries away from primary and extractive manufacturing to services represents a global trend, which is expected to continue into the future. Earlier it was stated that the concepts of JIT are being adapted to managing services. The following discussion addresses how JIT can be applied to managing the quality of services for logistics.

The meaning of services is difficult to define as a 'service' constitutes a large number of activities. To simplify the definition of a service, the activities pertinent to services can be grouped into three categories:

1. services provided directly to a customer in exchange for payment. Activities such as personal services from a medical doctor, dentist or hairstylist are included in this category;
2. services offered in support of a product, such as warranty, delivery of the product and other after-sale support activities;
3. services provided to a product, such as overhauls and repair work for heavy equipment, or products supplied by original equipment manufacturers.

Managing the quality of services, as it applies to the last two categories, is the focus of business logistics and is the subject of the following discussion. The quality of services provided is composed of several characteristics. Variability of the characteristics affects the degree to which a customer is satisfied with the service provided. The characteristics of services in the context of business logistics, as compared to physical products, which have important implications for managing quality include:

- Customer needs and performance standards are often difficult to identify and measure, primarily because customers define what they are, and each customer is different.
- The production of services typically requires a higher degree of customization than does manufacturing. Logistics managers must tailor their services to individual customers. In manufacturing, the goal is uniformity.

- The output of many service systems is intangible, whereas manufacturing produces tangible, visible products. Manufacturing quality can be assessed against firm design specifications, but service quality can only be assessed against customers' subjective expectations and past experience. Manufactured goods can be recalled or replaced by the manufacturer, but poor service can only be followed up by apologies and reparations.
- Services are produced and consumed simultaneously, whereas manufactured goods are produced prior to consumption. In addition, many services must be performed at the convenience of the customer. Therefore, services cannot be stored, inventoried or inspected prior to delivery as manufactured goods are. Much more attention must therefore be paid to training and building quality into the services as a means of quality assurance.
- Customers are often involved in the service process and present while it is being performed, whereas manufacturing is performed away from customers. For example, customers of a logistics service provider place their own orders for consignment, track and trace their cargoes with the firms and are regularly updated on the status of their cargo consignments.
- Services are generally labour intensive, whereas manufacturing is more capital intensive. The quality of human interaction is a vital factor for services that involve human contact. For example, the quality of logistics services depends heavily on the interactions among the shippers and the staff responsible for shipping documentation, cargo handling, warehousing and materials, trucking and freight forwarding and others. Hence, the behaviour and morale of service employees is critical in delivering a quality service.
- Many service organizations must handle very large numbers of customer interactions. For example, on a given business day, a shipping line might process many enquiries and orders for shipping spaces. A global logistics service provider might handle more than a million small parcels all over the world. Such a large transaction volume increases the opportunity for errors.
- These characteristics of services have made it difficult for firms to seek quality improvements for their logistics activities. Many firms have a quality management system for logistics management similar to that of their manufacturing counterparts, which tend to be more product oriented than service oriented. However, logistics activities have special requirements that manufacturing systems cannot fulfil. The most important dimensions of service quality for logistics activities include the following:
    - Time – how much time a customer is willing to wait?
    - Timeliness – will a logistics service be performed when promised?
    - Completeness – are all the items in the order included?
    - Courtesy – do frontline employees greet each customer cheerfully?
    - Consistency – are logistics services delivered in the same fashion for every customer, and every time for the same customer?
    - Accessibility and convenience – is the logistics service easy to obtain?
    - Accuracy – is the logistics service performed right the first time?
    - Responsiveness – can logistics service personnel react quickly and resolve unexpected problems?

Quality Issues in JIT

The previous discussions on quality addressed the increase in awareness of quality, the costs of building in quality and the guidelines to ensure quality. Developing methods to meet customer requirements effectively is what constitutes the very nature of JIT. The streamlining of operations, reduction of waste and quality improvements work together to attain the goals of JIT logistics more effectively. This section discusses the key practices that must be undertaken by firms before they embark on any quality improvement imitative for JIT logistics. These practices include CI, TEI, TQC, quality circles and the ISO standards.

CONTINUOUS IMPROVEMENT

Effectively communicating the meaning of TQC and TEI requires one to grasp a complete understanding of the concept of CI. CI is an important value statement that endorses the notion that a firm can continuously improve all its processes and activities through the application of systematic techniques. It also embraces the idea that there should be relentless, ongoing hunts to eliminate the sources of defects, inefficiencies and nonconformance to customer specifications, needs and expectations. Although CI is more readily recognized as being part of the philosophies and practices of manufacturers, its application need not be limited to the manufacturing context. It is also applicable for the management of business logistics.

Similar to JIT, CI operates according to two basic premises, namely to eliminate waste or non-value-adding activities, and to seek continuously to improve quality. To succeed, CI must become an integral part of a firm's philosophy, practices or even culture. The prerequisite to CI involves a change in attitudes. The current attitudes and beliefs held by many employees are continually reinforced through work practices by management, unions and shop-floor workers. Breaking down the barriers to effective communication between management and employees and replacing any adversarial bargaining systems with cooperative efforts lay the foundation for success.

The question remains: how does a firm attempt to implement CI? The ensuing discussion presents the elements of a general implementation framework:

- Changes are initiated from the top down – securing commitment from all will require management to set an example. Actions and effort must be consistent with the spoken word.
- Establish CI as a formal organizational policy – publicize the effort to inform customers and employees that CI is a means of achieving organizational goals.
- Involve the union in the process as soon as possible – union involvement at this stage is the first step towards creating a cooperative effort and provides the opportunity for union representatives to express their concerns openly.
- Listen to employees – communication should be from the bottom-up. Employee involvement at the earliest stages will reduce resistance, facilitate information flow and make employees an integral part of the process.
- Establish clear, objective goals for improvement – CI is an all-encompassing philosophy and must be made workable through the establishment of clearly defined and understandable goals. Failure to develop such goals will result in a frustrated

workforce, aware of the must for improvement, but lacking the focus and standards necessary to gauge their success.

TOTAL EMPLOYEE INVOLVEMENT

TEI is the collective efforts of all the employees directed towards problem solving and quality improvement. TEI differs from many of the traditional forms of employee involvement in that it seeks to involve all the employees. The basic underlying principle of TEI is that employees who work closest with the customer, the product and the processes are most able to recognize problems in these elements and know how to solve them.

TEI is realized through many methods used as a means of securing employee involvement. Suggestion systems represent one form of employee involvement. However, while they attempt to involve employees, they may fail in many respects because they are either not taken seriously or only a handful of employees will participate. The latter has been the main criticism of suggestion systems because many employees who have relevant suggestions fail to become involved in the process. The main advantage of using TEI is that it provides a simple method of keeping employees interested in their work through establishing employee ownership. Employees are asked to become involved in the operations of the firm in that they are invited to set its goals and ultimately its direction.

One important aspect of TEI is empowerment, which is the act of delegating responsibility and authority to employees to make decisions and to take action. Empowerment embraces the assumption that employees, when properly trained, can often make the best decisions on how to manage and improve a process. For empowerment to take hold, it requires a substantial and sustained commitment to a culture that values the contribution of all the employees and a management that will not and cannot rescind such authority once given to employees.

TEI suggests a culture characterized by open communication and the implementation of information technology, which makes information and the ability to consult with others easy and commonplace. To reduce the lead time required to fulfil customer requirements, it is highly desirable that the organizational structure possesses a cross-functional orientation, with concurrent information flows and decision making, and relatively few layers in the hierarchy (Koufteros et al. 2007a). It is imperative to have workers that can plan and execute their own work, and promote a management style that emphasizes collaboration and consensus. The following provide a list of practices for managers who seek to make empowerment an important part of TEI:

- Foster ownership – this suggests that management has recognized that employees who have the responsibility for processes, projects and tasks must also have the authority to exercise this responsibility.
- Value all the contributions – it is important to appreciate that everyone in the firm has something to contribute, and to create an environment where each employee is free to make their contribution.
- Listen to the smallest voice – you never know from whom the answer to a problem might come. Again, create an environment where this is a core value.

- Allow teams to own problems – management must give teams autonomy to investigate and solve the problems they have taken on. If this does not happen, then there is no reason to create such teams.
- Delegate authority to the lowest possible organizational level – managers should always work at giving the people who have been hired to do jobs the authority and responsibility to do those jobs (and encourage learning from mistakes).

As firms shift from an industrial to a post-industrial mode of operations that embraces the JIT principles, they need a structure that has: 1) rules and regulations that encourage creative, autonomous work and learning (the nature of formalization); 2) few layers in the organizational hierarchy to enable quick response; 3) a high level of horizontal integration to increase knowledge transfer; 4) a decentralized decision-making process so operating issues can be dealt with effectively an quickly; and 5) a high level of vertical and horizontal communication to ensure coordinated action (Koufteros et al. 2007b).

Total quality control

The Japanese interpretation of TQC involves removing quality control from the sole responsibility of designated specialists to all the disciplines in a firm. Thus, everyone in the firm would have a stake in managing quality. TQC is a philosophy aimed at increasing customer satisfaction through continuous process improvements. The customer, as it applies to this definition, can be either internal or external to the firm. TQC is fundamental to JIT as it functions to eliminate the 'cause' of quality problems.

TQC contains three important elements, which can be understood through analyzing the individual meaning of 'total', 'quality' and 'control'. Quality is about meeting customer requirements. Total quality denotes that everyone within a firm is responsible for quality improvements. Thus, the finance, marketing, manufacturing, personnel, engineering, production and materials personnel will all play a role in achieving quality. TQC requires the involvement of all employees and is concerned with their ability to affect the processes, and product and experience results to meet established standards.

The TQC process operates according to four basic principles in the Deming's Plan-Do-Check-Act (PDCA) cycle to deal with quality problems. The PDCA cycle for quality improvement is provided in Figure 5.1.

1. Plan – this involves identifying, prioritizing and assessing the impact of a problem. Assigning the problem to a group or individual to ensure it is rectified is fundamental to this stage.
2. Do – the second component of the cycle requires assigning a cause or causes to the problem and identifying symptoms.
3. Check – this involves assessing the effectiveness of the solution. Failure to experience improvements indicates that either the cause or the effect of the problem was incorrectly identified. Counter-measures to the solution must also be checked. Thus, in the event that the solution proves to be ineffective, efforts must be directed once again to 'do'.

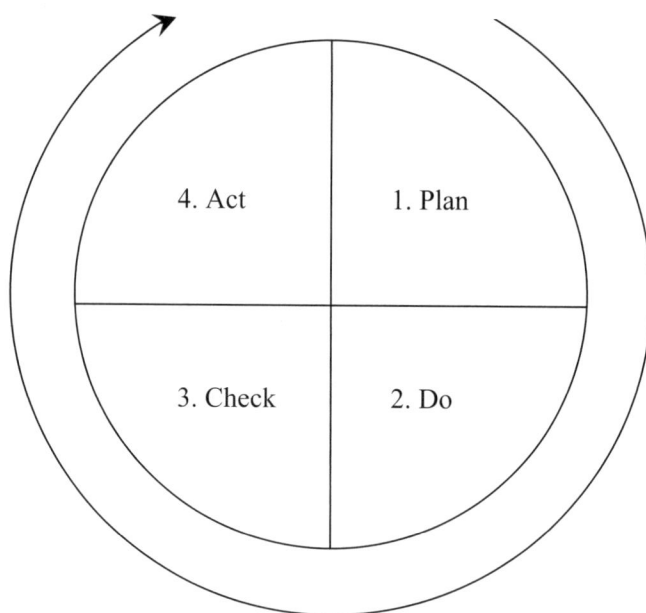

Figure 5.1 PDCA cycle of improvement

4. Act – the realization of an effective solution should lead employees to standardize the improvement by first identifying what other areas can benefit from this change. Implementation of counter-measures can be done in other areas of the plant that are likely to experience the problem.

The PDCA procedure of TQC demonstrates how problems are detected, prevented and corrected, and how operational standards are established. In the 'plan' stage, problems are identified, prioritized and analyzed to assess their impact. The problems are then assigned to respective personnel for rectification. In the 'do' stage, root causes and symptoms of the problems are identified. During the 'check' stage, solutions to the problems are evaluated in order to ensure that the causes and symptoms of the problems are correctly identified. Finally, in the 'act' stage, employees assess how other areas can benefit from the developed solutions and these solutions evolve as standardized procedures. These principles are designed to make everyone conscious of the importance of improvement and learning from experience. The PDCA cycle suggests that managers, teams and all the employees plan improvements in a process, implement them (sometimes in a pilot programme), check how well they are working, make modifications if appropriate and then implement them as standard policy for that and other related processes.

QUALITY CIRCLES

Quality circles are teams of staff who are volunteers. The team selects issues or areas to investigate for improvement. To work properly teams have to be trained, first in how to work as a team and secondly in problem-solving techniques. Quality circles are identifiable through the purpose and action of members, as well as by several unique characteristics that they possess. The following provide the principles for a quality circle to work effectively:

- The circle should only consist of volunteers.
- The members of the circle should all be from different functional areas.
- The problem to be studied should be chosen by the team, and not imposed by management. The problems looked at by the circle may not always be directly related to quality or, initially, be seen as important by management.
- Management must wholeheartedly support the circle, even when initially decisions and recommendations made by the circle are of an apparently trivial nature or could cost the company a hefty sum to implement (such as a recommendation for monogrammed overalls).
- The members of the circle need to be trained in working as a team and problem-solving techniques, and in how to present reports. The basic method study approach of asking why (what, where, when, who and how) is a standard quality circle approach to problem solving and members will need to be taught how to apply this structured approach to solving problems.
- The leader of the circle and internal management of the circle should be decided by the members.
- Management should provide a middle manager as a mentor to the circle. The mentor's role is to assist when requested and generally to provide support. The mentor does not manage the circle.

The overall tenor of these principles is trust and employee involvement. Management of the firm has to be seen to be willing to trust the members of the circle to act responsibly and then must be active in supporting the circle. Although, initially, the circle may not appear to be addressing hard quality issues, as the confidence of the members increases, very real benefits can be expected. Side benefits of quality circles, but nonetheless important, are the fostering of a supportive environment that encourages workers to become involved in increasing quality and productivity, and the development of problem-solving and reporting skills of lower-level staff.

The benefits derived from the efforts of quality circles are numerous. It has been established that they contribute to higher levels of efficiency, quality, productivity, safety, cost effectiveness and lower absenteeism. However, the main benefit of quality circles is not the cost savings that can be experienced by the implementation of the solutions. The main benefits of quality circles are:

- increased employee commitment through participation and consensus decision making;
- increased sense of employee ownership and control;
- providing a mechanism to realize employee potential;
- improved communication between employees;
- increased motivation and employee morale;
- an effective means of training.

The potential benefits a firm can experience through the use of quality circles outweigh the costs. However, an awareness of the costs associated with quality circles will provide a realistic picture of what quality circles are capable of accomplishing. The most effective approach to implementing quality circles is to start with a realistic attitude. Quality circles do not provide quick-fix solutions to all organizational problems. Many problems will remain outside the responsibility of the circle members. Common costs and obstacles associated with the implementation of quality circles include the following:

- The cost of training employees.
- Union concerns – unions may hold some reservations to management regarding the implementation of quality circles and their use. Objections may include the following:
  - the use of quality circles as a means of eventually eliminating the union;
  - increasing productivity levels to the point where it will produce adverse effects on the workforce;
  - reducing the size of the workforce through increased productivity levels.

The most effective approach to avoid union attempts to obstruct implementation efforts is to maintain open communication with union representatives. Sharing the objectives of the quality circles will assist in clarifying management motivation.

- Quality circles remove the worker from the job, reducing the time required to complete the tasks. The experience of most firms with quality circles is that workers are able to accomplish the same volume of work while participating in weekly meetings. This is a plausible outcome as the object of quality circles is to develop more effective approaches to task completion.
- Resistance to change – the implementation of quality circles represents a change to established work methods. Resistance is likely to be directed at management from supervisors and employees. Common concerns frequently address the increased level of responsibility of workers and reduced supervisory control.
- Failure with previous attempts to increase the level of employee involvement. Past attempts may have resulted in failure, however, this need not be representative of the outcomes of quality circles. The success with quality circle implementation will require a positive attitude, commitment from all involved and a well-defined plan of action.
- Failure to provide sufficient training to members of the circles – this will result in incorrectly identifying cause, effect and formulation of solutions. The circle will lack the direction required to function constructively if the circle leader is not provided with adequate training.
- Lack of management support – failure of management to demonstrate commitment to the circles will result in failure. Circle members need management's support from provision of the necessary tools for problem solving to feedback on suggestions.
- Setting unrealistic expectations for quality circles – the most effective approach to managing quality circles is to allow them necessary time to develop and grow on their own. Expecting too much too soon will result in groups of frustrated and defeated employees.
- Failure to establish specific, quantitative measures.

Successful quality circles contain all the features that are known to contribute to their success. Failure to address at least one of the obstacles to their implementation will interfere with the effectiveness of their long-term functioning.

ISO 9000

The ISO 9000 series and the more recent 14000 environmental series have been developed over a long period of time. ISO 9000 is a set of quality assurance standards that was published by the International Organization of Standardization (ISO) in 1987. The ISO was founded in Geneva, Switzerland in 1946 to develop international industrial standards. The ISO 9000 standard is codified, verifiable and easily adaptable. In fact, it is so adaptable that updates and changes have been made approximately every 3 years since its inception, with the last significant update made in 2000.

ISO 9000 certification means that a firm constantly meets rigorous standards, which are well documented, of managing the quality of its products and services. To retain certification, the firm is audited annually by an outside accredited body. ISO on the letterhead of a firm demonstrates to itself, to its customers and to other interested bodies that it has an effective quality assurance system in place. In sum, ISO 9000 serves to give customers confidence that the product/service being provided will meet certain specified standards of performance and that the product/service will always be consistent with those standards. Often, some customers will insist their suppliers have ISO accreditation.

There are also internal as well as strategic benefits for firms that seek to become ISO 9000 registered. The internal benefits include requiring that all the business activities related to a product be conducted in a three-part continuous cycle of planning, control and documentation. This cycle, as well as a necessary quality conformance system that maintains regular calibration of measuring and testing equipment, are some of the immediate benefits. These also prevent the shipment of products that do not meet quality standards, thereby reducing the number of returns from buyers. The strategic benefits include gaining access to companies that only deal with ISO certified suppliers. These companies wish to maintain high quality levels while keeping costs low. One way of doing this is to have their suppliers examine quality first. By doing so, the buyer may reduce the number of inspections of incoming materials.

To attain accreditation, a firm has to prove that every step of the process is documented and that the specifications and check procedures shown in the documentation are always complied with. The recording and documenting of each step is a long and tedious job. Perhaps the most difficult stage is agreeing on what are considered standard procedures to be documented.

If a firm does not have a standard way of doing things, trying to document procedures will prove difficult and many interesting facts will emerge. The act of recording exactly what is happening and then determining what one set method should be will in itself be a useful exercise. Non-value-adding activities should be unearthed, and hopefully, overall, more efficient methods will emerge and be adopted as standard procedures. Determining a standard does not imply that the most efficient method is being used. The standard adopted only means that there is now a standard method (not necessarily the most efficient), that the method is recorded and that the recorded method will be used every time. The standard method not only includes the steps taken in the process but lists the checks and tests that will be carried out as part of the process. This will often require the design of new and increased check procedures, and a method of recording that each check or test has been done.

Nevertheless, there is considerable debate over the value of the ISO 9000 certification process and a number of criticisms of ISO 9000 have appeared. These are:

- it does not ensure product quality;
- it is seen as an end in itself rather than the beginning of a quality journey;
- it is driven by documentation not organizational behaviour;
- it is bureaucratic, stifles change and quickly becomes out of date;
- it is limited to those parts of the organization that directly affect the quality of the product or service and not the whole of the operation;
- it focuses on technologies, methods and systems, not on the competencies and skills of people and their creativity;
- it is often an imposed system that diminishes ownership and hence motivation; it involves following procedures rather than taking responsibility;
- it under-emphasizes improvement;
- it does not add value.

The claimed advantages and criticisms of accreditation have been previously presented in dualistic terms. Which side one comes down on seems to depend rather heavily on how the process is implemented. For example, management may see accreditation as an end in itself rather than the beginning of a quality journey, and may be driven by procedures and documentation rather than organization behaviour. If so, all one will get is transactional change and perhaps window dressing. However, if management emphasizes long-term capability, rather than short-term gains, then one would expect to see accompanying transformational change.

INNOVATION AND KNOWLEDGE MANAGEMENT

In the face of intensifying global competition and an exponential explosion of knowledge, firms around the world striving to maintain their competitiveness in the marketplace need to strengthen their ability to deploy and exploit their tangible and intangible organizational assets to the fullest (Yeung et al. 2007). Increasingly, the ability of firms to achieve their logistics goals and create customer value relies on their success in managing such intangible assets as human skills, knowledge bases or other strengths that competitors cannot provide.

Cohen and Levinthal (1990) argued that the ability of firms to recognize the value of new, external information, assimilate it and apply it to commercial ends is critical to its innovative capacity. Innovation requires the ability of firms to translate and exploit knowledge into the discontinuous aspects of social, sectoral, technological and market development, where knowledge management is a critical constituent in the process. Knowledge management not only better enables firms to accomplish within-paradigm improvements such as CI, but also paradigm shift, that is, breakthrough innovations (Baker and Sinkula 1999). It refers to the organizational process that is concerned with the creation, storage, retrieval and application of knowledge. To create customer value, it is highly desirable for firms to innovate and develop knowledge stores related to the desirability of prospects, customer defection intentions, needs and preferences of customers, likely profitability of current and prospective customers and emergence of market threats (Massey et al. 2001).

The relationships among innovation, knowledge management and performance in business logistics can be examined from the Resource-Based View (RBV) of the firm (McEvily and Chakravarthy 2002). The RBV defines strategic assets as rare, valuable, imperfectly imitable and non-substitutable resources that are tied semi-permanently to a firm (Wernerfelt 1984) and enhance the organizational capabilities of the firm, such as management systems and technical knowledge. This knowledge, for example, business logistics, thus becomes a strategic asset that is the source for creating sustainable competitive advantages. Such firms are viewed as a social institution whose knowledge is stored in its rules or behaviours, which are constantly being shaped, preserved and improved (Hult and Ketchen 2001). The field of accounting offers a similar perspective on the relationship between knowledge management and business performance. The Balanced Scorecard Strategy Map (BSSM) developed by Kaplan and Norton (1992) provides a conceptual foundation for such a perspective. The BSSM sheds light on the knowledge, skills and system that employees in an organization will need – learning and growth – to innovate and build the right strategic capabilities and enhance efficiency. These efficient internal processes deliver specific value to the market – the customers – and eventually lead to higher shareholder value – the financials. It embraces the flows of value from employees to customers and finally to investors. The BSSM offers valuable insights for firms to manage the creation process of logistics activities from employees through to the investors.

Quality Management Tools

Quality management tools are essential ingredients and instrumental to the success of quality improvement efforts for JIT logistics. Although these problem-solving tools are not a panacea for quality problems, they can be effective means for solving them. The problem-solving tools used by quality circle members can be applied to solving quality problems and addressing organization-wide quality issues. These problem-solving tools include:

- methods for identifying causes and effects such as brainstorming and the nominal group technique (NGT);
- the 5 'whys';
- Pareto analysis;
- histograms;
- check sheets;
- flip charts;
- cause and effect or Ishikawa diagrams;
- benchmarking.

BRAINSTORMING

Brainstorming can be considered as a relaxing means for identifying all the causes of a problem. It consists of a group of people being given a problem to consider with every person encouraged to make at least one suggestion. When solutions to the problem cannot be easily formulated or there are just no clues to begin, brainstorming is always a good way to start. It allows free expression of ideas and thereby stimulates creative thinking. Brainstorming works well in groups. In order to make the process a success, it is important to create a positive atmosphere that encourages active participation.

Before the actual brainstorm process begins, it is important that the subject be defined and that the rules of the session are agreed. Members of the team will need at least 5 minutes of thinking time before the brainstorming proper begins. Following are some rules for successful brainstorming:

- one person is responsible for recording suggestions on a white board or a large flip chart;
- encourage everyone in the team to 'freewheel'. There should be no criticism of seemingly silly suggestions;
- everyone in the team should come up with at least one suggestion, and other members of the team should not interrupt or make comments;
- take suggestions by working around the room so that everyone has a turn;
- if someone is unable to contribute first time round, pass on to the next person;
- typically, there will be a lot of suggestions in the first 20 minutes, then there will be a lull. Don't stop when this lull occurs but keep going, as usually there will then be another burst of ideas. Often the second burst provides the most creative ideas;
- keep the initial ideas in front of the team until the end of the brainstorming session;
- when suggestions have dried up, the team should review the suggestions made and sort them into logical groups. Some suggestions will be found to be duplications and can be eliminated. One method of sorting the suggestions is to use a form of the cause and effect diagram.

THE 5 WHYS

The 5 'whys' can be applied to any problem to determine its real cause. The underlying purpose of asking 'why' five times is to avoid attributing a false cause to the problem. The tendency of most people is to assign the first symptom or cause identified to be the root of the problem, whereas there may be several causes or symptoms to a problem. It is a variation on the classic work study approach of 'critical examination', involving six questions: why, what, where, when, who and how? The objective is to eliminate the root cause rather than the patch up effects. The 5-why process requires an employee to consider all the factors that may be part of the problem. Factors to consider include the tools and equipment used in production, work practices, the flow of work and work-in-progress. The use of 5 whys involves the following steps:

- select the problem for analysis;
- ask five 'close' questions, one after another, starting with why;
- do not defend the answer or point the finger of blame at others;
- determine the root cause of the problem.

PARETO CHART

A Pareto chart is a special form of bar chart that rank-orders the bars from highest to lowest in order to prioritize problems of any nature. A Pareto chart assists in problem identification and analysis of the percentage of each cause that contributes to the problem. It is known as 'Pareto' after the nineteenth century Italian economist Wilfredo Pareto, who observed that 80 per cent of the effects are caused by 20 per cent of the causes, that is, 'the 80/20' rule. The 80/20 rule, however, need not apply to all situations. For example, after careful analysis, it may be determined that 75 per cent of the problem is caused by 25 per cent of the causes. The following steps apply for the preparation of a Pareto chart:

- select the subject for the chart, for example, a particular product line exhibiting problems, or a department, or a process;
- determine what data need to be gathered. Determine whether numbers, percentages or costs are going to be tracked. Determine which nonconformities or defects will be tracked;
- gather data related to the quality problem. Be sure that the time period during which data will be gathered is established;
- use a check sheet to gather data. Record the total number in each category. Categories will be the types of defects or nonconformities;
- determine the total number of nonconformities, and calculate the percentage of the total in each category;
- determine the costs associated with the nonconformities or defects;
- select the scales for the chart. The vertical axis scale is typically the number of occurrences, number of defects, dollar loss per category or per cent. The horizontal axis usually displays the categories of nonconformities, defects or items of interest;
- draw a Pareto chart by rank-ordering the data from the largest category to the smallest. Include all pertinent information on the chart;
- analyze the chart or charts. The largest bars represent the vital few problems. If there do not appear to be one or two major problems, re-check the categories to determine whether another analysis is necessary.

HISTOGRAMS

Histograms are a graphical representation of recorded values in a data set according to frequency of occurrence. They are bar charts of numerical variables giving a graphical representation of how the data are distributed. When measurements are taken from a process, they can be summarized by the use of a histogram. Data are organized in a histogram to allow those investigating the process to see any patterns in the data that would be difficult to see in a simple table of numbers. The data are distributed into classes in the histogram. Each interval on a histogram shows the total number of observations made in each separate class. Histograms display the variations present in a set of data taken from a process. There are several advantages of applying histograms in CI projects including:

- they display a large amount of data that are difficult to interpret in tabular form;
- they illustrate quickly the underlying distribution data, revealing the central tendency and variability of a data set.

CHECK SHEETS

Check sheets offer an efficient method of collecting data that can be applied to any problem and is useful in illustrating patterns or trends. Check sheets are tables used to tally the frequency of defects or problems that have occurred. These sheets allow a team to systematically record and compile data from observations so that trends can be shown clearly. Constructing a check sheet should entail the following steps:

- agree on the type of data to be recorded. The data could relate to the number of defects and type of defects and apply to equipment, operator, process, department, shift and so on;
- decide which characteristics and items are to be checked;
- determine the type of check sheet to use, for example, tabular form, defect position or tally chart;
- design the form to allow the data to be recorded in a flexible and meaningful way;
- decide who will collect data, over which period, and form what sources;
- record the data on check sheets and analyze the data.

FIVE S's

The five S's are a tool for improving the housekeeping of an operation, developed in Japan, where the five S's represent five Japanese words all beginning with 'S', which are:

- Seiri (structurize) – means organizing and throwing away things you don't use.
- Seiton (systematize) – stands for neatness in the workplace.
- Seiso (sanitize) – applies to cleaning.
- Seiketsu (standardize) – stands for standardization.
- Shitsuke (self-discipline) – refers to the discipline required to maintain the changes that have been made using the first four Ss.

The five S's method is a structured sequential programme to improve workplace organization and standardization. Five S's improves the safety, efficiency and the orderliness of processes and establishes a sense of ownership within the team. In sum, the five S's are useful to help start quality initiatives because they help develop the discipline needed to improve quality.

ANALYSIS OF CAUSE AND EFFECT

This is a method used to display and study underlying causes. Some people called this 'fishboning' since the output is a chart that takes the shape of a fish skeleton. An example is shown in Figure 5.2. The analysis was first proposed by Kaoru Ishikawa, a Japanese pioneer in quality control methodologies. The purpose of the diagram is to assist in brainstorming and enable a team to identify and graphically display, in increasing detail, the root causes of a problem. The following figure illustrates the generic format for a cause-and-effect diagram.

Quality Management Issues 133

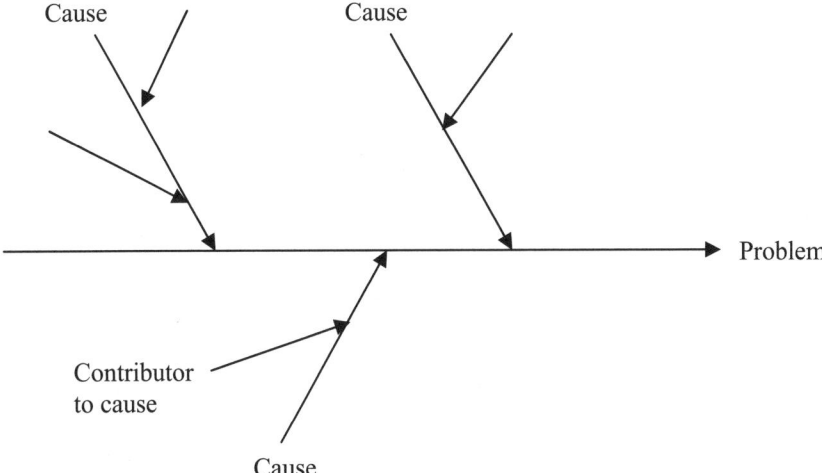

Figure 5.2 Cause-and-effect diagram

Constructing a cause-and-effect diagram involves the following steps:

1. Clearly identify the effect of the problem. Place the succinctly stated effect or problem statement in a box at the end of a horizontal line.
2. Identify the causes. Discussion ensues concerning the potential causes of the problem. To guide the discussion, attack just one possible cause area at a time. General topic areas are usually methods, materials, machines, people, environment and information, although other areas can be added as needed. Under each major area, sub-causes related to the major cause should be identified. Brianstorming is the usual method for identifying these causes.
3. Build the diagram. Organize the causes and sub-causes in a diagram format.
4. Analyze the problem. At this point, solutions will need to be identified. Decisions will also need to be made concerning the cost-effectiveness of the solution, as well as its feasibility.

BENCHMARKING

Benchmarking is the act of continually comparing the performance of one process against that of another, usually of comparable or greater performance and frequently on a continuous basis. The goal of benchmarking is to identify best practices and then adapt those to improve one's own processes. Most benchmarking projects, when properly done, follow a highly structured set of steps. These steps include:

- identifying precisely what you will benchmark;.
- creating a list of benchmarking candidates;
- comparing data between your business activity and the benchmark;
- establishing goals and action plans for improvement based on what you have learned from undertaking this project.

The benchmarking technique has delivered many tangible and intangible benefits, particularly it:

- enables process owners to identify what is to be improved and motivates them by the knowledge of what is achievable;
- enhances communication, trust and partnership spirit between participating units by focusing on the high visibility of key processes and performance indicators;
- removes complacency even for high performers by providing a platform for improvement;
- focuses on processes rather than individuals and helps to eliminate defensive practices;
- focuses attention on the details of best practice, thereby initiating a process of generating a learning culture;
- has a remarkable effect on strategy formulation, strategy implementation and leadership development.

The principle behind benchmarking is that to learn how well anything works, you need something to which you can compare it. Finding outstanding firms and understanding how they perform provides guidelines for process and organizational improvement. It is useful to note that outstanding performers in a particular process do not have to be in the same industry. What is important to consider is the process itself and not necessarily the industry or firm in which it is undertaken.

Summary

In this chapter we discussed the quality management issues for the implementation of JIT logistics. Quality improvement is critical to the success of JIT logistics. Quality should be built into the design of logistics activities before they are performed. The costs of quality should not be overlooked as the costs of poor quality can often be substantial. It is important that the quality issues of the logistics activities performed should be well addressed by a firm since they make a direct impact on customers' perceptions of the logistics service quality of the firm. Firms can ensure that they are continuously committed to the notion of 'quality improvement' by implementing various quality management practices and tools for performance improvements.

CHAPTER 6
Implementation Issues

The implementation of JIT logistics requires proper understanding, planning and organizing of the related activities, with the development of a strategy and operational plan to begin with. The implementation of JIT logistics is an incremental process, which needs continual efforts over an extended period that may run into years to realize the full benefits. It should be noted that the implementation process is likely to invoke changes in a firm. An understanding of the organizational factors is essential to avoid potential pitfalls in the implementation process. The development of an implementation strategy and an operational plan will provide a roadmap necessary to kick-off and conclude JIT logistics implementation.

Organizational Factors for Implementation

The implementation of JIT logistics is a challenge in itself. Organizational adjustments, the introduction of new forms of management, decision-making patterns, work assignments and reward systems are some of the modifications that can occur within organizations. Several studies have shown that a successful JIT implementation is a complex process characterized by:

- an evolution of the organizational culture in place and a revision of the working methods and procedures;
- the implementation of new administrative procedures;
- the necessity of not perturbing or confusing the normal operations of the firm through poor human resources implementation;
- the necessity of respecting the market, and the financial and information technology positions of the firm;
- acknowledging the precedence of some steps over some others.

Furthermore, the implementation process implies a different attitude towards customers and suppliers. The focus is not on price and cost savings for one's own company, but finding mutually beneficial solutions for all the parties by fully utilizing the competencies available in the logistics activities. This process includes determining which customers and suppliers should be in focus through closer cooperation. This closer cooperation increases awareness of the resources expended on each relationship.

The benefits of JIT do not just happen. Before a firm can enjoy any benefits from implementing JIT logistics, it must accept JIT as an organizational philosophy. This requires the organization to change or modify its operating procedures, logistics systems and, in most cases, its organizational culture. Resources for cost and service improvements have to be committed, relationships with suppliers and customers have to be re-defined, quality circles have to be established and accurate demand forecast has to be achieved and maintained. To succeed in the implementation of JIT logistics, it is important for firms to consider the following organizational factors:

- top management support;
- employee involvement;
- information sharing.

TOP MANAGEMENT SUPPORT

Top management support refers to the commitment and personal involvement of senior management in the implementation of JIT logistics. Top management must make JIT a priority and communicate its importance to all the employees. Because JIT affects many corners of the firm, top management must obtain support from all the organizational functions and all the management levels (Beer 2003). Their roles can include personal participation in the implementation committee, process redesign, creation of goals, measures and policies, providing training on JIT skills and knowledge, and allocation of appropriate resources. Other activities include participating in process teams, monitoring of progress, rewarding those who have contributed to cost reduction and service improvement in the implementation process.

The support of top management for the implementation of JIT is vital to establish credibility of the concept within the firm, assure continuity and enhance the chances of long-term success. The senior management team can make JIT the highest organizational priority and the top item on the management agenda. They must have faith in the long-term plan for JIT logistics and not expect immediate financial gains from the implementation. Ultimately, they are responsible for the organizational environment, behaviour, values, climate and style of management with which the implementation of JIT logistics will either flourish or wither. Top management needs to create and promote an environment in which, for example:

- people can work together as a team;
- teams work with teams;
- mistakes are freely admitted without recriminations, and anything that occurs is perceived as an improvement opportunity, that is, a no-blame culture;
- people are empowered to make decisions;
- people improve on a continuous basis in the processes under their control;
- people direct their attention to identifying, satisfying, delighting and winning over customers, whether they are internal or external;
- ideas are actively sought from everyone;
- development of people is a priority;
- employee involvement is worked at continually;
- permanent solutions are found to problems;
- departmental boundaries between functions do not exist;
- effective two-way communication is in place;
- recognition is given for improvement outcomes.

The idea of top management support is vital to the success of JIT logistics implementation. Many JIT implementation problems are related to lack of top management support and resources committed to employee involvement and training. This is because top management creates the environment to which everyone else in the firm adapts. If firms' leaders suggest that JIT is a good idea and then assign its implementation to middle managers while not changing their own values and behaviour and providing the needed resources, JIT will not happen in those firms. JIT implementation involves a company-wide change and requires that everyone in a firm must be involved in activities of reducing wastes and improving services. The process starts with top management support. Top management should provide the needed support and resources to initiate the change process, encourage employees to participate, equip them with the needed knowledge and skills, and reward them for achievement in cost reduction and service improvement.

The implementation process also initiates a change process in the organization as every member of the firm will be affected somehow. To reduce resistance towards the change, management will need to demonstrate its commitment to JIT and develop a trusting relationship between managers and workers. Only top management can bring about such cultural change by committing to the value behind such changes. Another important role of top management is to make this new culture of JIT not dependent on anyone or a few managers. This goal should be for the culture to become so strong that it is perpetuated by everyone in the firm.

EMPLOYEE INVOLVEMENT

Employees represent the core of a firm's capabilities because they provide the intellect, empathy and ability that are required to reduce wastes and improve services. Resistance from employees to participate in the change process may cause the implementation of JIT to fail. Therefore, a mechanism must be in place to develop, train, care for and motivate people to achieve the JIT goals.

Employee involvement means deploying management practices that encourage and enable employees to routinely play a critical role in making decisions about operations, setting priorities, making and implementing suggestions for improvements, planning, setting goals and targets, and monitoring performance. The idea is similar to that of employee empowerment. To this end, senior management needs to get employees more involved in making day-to-day decisions that are usually made by middle management.

Employees who are to be involved in the implementation of JIT logistics must be clear about and conscious of the processes of change that are occurring internally within the firm. It is not uncommon that unchanged management and employee behaviours are barriers to CI in the implementation process. As a result, the focus should not only be on identifying improvement opportunities, but also on change management and reward systems, which encourage behavioural changes towards attaining the JIT goals.

Germain and Droge (1997) argued that as the extent of JIT implementation increases, employees are more likely to identify problems and suggest solutions that enhance the firm's performance. Furthermore, Jayaram and Vickery (1998) suggested that employee empowerment is related to procurement lead-time performance. Teams comprised of members from purchasing, materials control, process/design engineering, production and suppliers can solve many quality problems from supplies and enable the firm to better meet the market requirements. It is important to clearly communicate an understanding of the JIT concept, its conditions and the breath of implementation the firm wishes to engage in. The first stage of this communication process should encompass the employees who will be directly involved in building relationships with customers and suppliers. Next, the employees who are involved via their role in internal processes, as well as those in the firm at large, should be brought into the loop of understanding.

The lack of JIT training, education and expertise has been identified as one of the critical problems encountered in the implementation of the JIT management approach (Salaheldin 2005). One way of educating employees about the concept of JIT logistics is to have the responsible pioneer manager teach courses on JIT logistics in the firm. This method also allows employees performing different functions to share their knowledge with one another, thereby supporting the firm's orientation towards cost and service improvement in logistics activities. The same form can be used during the start-up phase of a JIT project, where there will often be participants in sub-projects, who are less aware of the concept of JIT logistics, but who nonetheless play a central role in one of the processes important to the firm involved in the implementation of JIT logistics.

The central competences that the implementation of JIT logistics must develop are the ability and desire to challenge the existing structure and ways of doing business in order to generate improvements in the logistics activities through openness, shared understanding of the processes and mutual respect for the benefit of all the involved parties.

There is a need to understand the employee-related issue when considering a switch from a traditional to a JIT approach to logistics management. For example, in a traditional manufacturing plant, workers are kept busy by letting machines run, since an idle worker or an idle machine is considered a waste. On the contrary, JIT views keeping a machine running just for eliminating idle men or machine hours as a waste, since JIT emphasizes the notion that nothing should be produced until there is a need from the downstream operation. As a result, idle workers are kept busy by running several machines, since each worker is trained to a level where they can perform multiple jobs. Such changes may create anxiety to employees due to the potential loss of their jobs. Employees must understand that they are integral to that system. They must be made to feel important and necessary for the continued survival and growth of the firm. This requires redefined jobs, possibly better pay and more training for employee involvement to be effective. The following is a list of practices that are useful for managers seeking to enable employee involvement in the implementation of JIT logistics:

- Foster ownership – this suggests that management has recognized that employees who have responsibility for processes, projects and tasks must also have the authority to exercise this responsibility.
- Value all the contributions – it is important to appreciate that everyone in the firm has something to contribute, and to create an environment in which each employee feels free to make their contributions;
- Listen to the smallest voice – you never know from whom the answer to a problem might come. Again, create an environment in which this is a core value.
- Allow teams to own problems – management must give teams autonomy to investigate and solve the problems they have taken on. If this does not happen, then there is no reason to create such teams.
- Delegate authority to the lowest possible organizational level – managers should always work at giving people who have been hired to do jobs the authority and responsibility to do those jobs.

INFORMATION SHARING

At the strategic level, logistics is concerned with the strategic choice of partner firms to facilitate the five critical flows in a supply chain, namely information, product, service, finance and knowledge flows. At the operational level, it involves inter-organizational processes that span organizational functions within the firm and link those of partner firms in the supply chain, which are deployed to coordinate and manage these five critical flows (Lai et al. 2004). Due to the trends of outsourcing non-core activities and global sourcing, business firms are increasingly transacting and communicating with a large number of external parties, for example, suppliers and distributors, to deliver products and services that meet the needs of the market (Mentzer et al. 2001). The integration of physical processes and the interchange of information are becoming essential to the management of business logistics. To integrate inter-organizational processes in supply chains for better coordination and mutual benefits, information technologies are often applied between partner firms that are willing to commit to the supply chain and to exchange information electronically. The growing deployment of information technologies allows firms in the supply chain to reduce financial risks, lower costs and improve quality by focusing on their core businesses, resulting in fewer relations to manage. Moreover, better economies of scale, a higher utilization of capacity and the formulation of more effective long-term plans can be achieved through the sharing of information between organizations and with an integrated supply chain. These strategic and operational benefits can translate into competitiveness in costs and services for all the partner firms in the supply chain.

There is a growing recognition that firms on their own cannot maintain all the necessary capabilities to compete. To do this effectively, firms must be able to develop partnerships with their suppliers and customers. It is by this extension of the logistics processes that better decisions are made possible, in terms of procurement, production, inventory and fulfilment. If a firm is to make effective logistics decisions, then it needs to ensure that there is an accurate, timely flow of information across the supply chain as a whole. If firms are to achieve this, they need to ensure that they have a set of systems that are capable of viewing the entire logistics process. This would include everything from supplier inventory positioning through market demand forecast to delivery modes.

Today, many firms have implemented ERP systems such as SAP, PeopleSoft, Oracle and BAAN. These systems make it possible to integrate activities and processes within individual firms, and offer the potential to transfer information to other actors via EDI/Internet/Intranet. They also facilitate frequent and reciprocal information exchange between actors regarding inventory status, forecasts, production plans, sales and marketing strategies and so on. The goal is to reduce uncertainty and reaction time for the logistics activities with the information readily accessible to all the parties as it is for any single participant in the implementation of JIT logistics. The emphasis should be on ensuring transparency in the logistics activities so that individual participants can plan their own activities based on information that is as reliable as possible.

The advances in information technology not only enhances the logistics processes by enabling instant communication and improving the accuracy of operational details, but also allows extensive operational data to be collected and sophisticated analysis on the gathered data to be performed. Developments in the electronic era are making organizational boundaries more fluid and encouraging a freer flow of information.

Channel members are also able to share and analyze their collective operational details more and more easily.

Given the importance of information sharing for the implementation of JIT logistics, there are several issues that need to be overcome. Lee and Billington (1992) presented a number of pitfalls in SCM that fall into three major areas, namely information definition and SCM, operational problems, and strategic and design related. The firm implementing JIT logistics also needs to be aware of the importance of defining information and its accuracy and establishing performance metrics. It requires changing the way in which firms do business and firms must be willing to:

- openly share information with suppliers and customers;
- create horizontal business processes;
- rely on a smaller number of outside suppliers;
- increase organizational and process flexibility;
- coordinate processes across organizational boundaries;
- empower employees to make process decisions;
- make real-time decision supply systems available.

These changes in business practices are not easy for firms to implement. Information is generally viewed as providing an advantage over competitors and firms resist sharing with their own partners. A major barrier to information sharing is that, while firms readily seek improved business performance, they are less willing to share risk. Firms are traditionally structured along functional boundaries and not aligned with business processes. Information sharing between partner firms may take a considerable amount of time to develop and nurture. However, achieving the benefits of these partnerships relies on the continuous flow of materials and information along the supply chain. It is felt that such collaboration can only be successful if there is effective communication between partners in the supply chain. Suppliers must have easy access to buyers' demand and real-time operational information, and suppliers must share critical scheduling information with buyers. Sharing of such information helps to reduce problems related to the timing, quality and quantities of deliveries.

Trust between the involved parties is one of the cornerstones for the implementation of JIT logistics. Increased understanding and respect for others' businesses is one way trust can be built. This process includes an awareness of a shared mission, vision and values for all the cooperative partners in the supply chain. A common understanding of what motivates decisions, decision-making processes and what defines success for each party is key to the cooperative effort for the successful implementation of JIT logistics. Mutual understanding of the expectation for the cooperative efforts and knowledge of the business areas of the firms involved increases the ability of the cooperative partners to act based on the common goals for JIT logistics. As trust can only develop over time, the cooperating parties must have the possibility of working together for a longer period of time. Trust can be demonstrated by informing one another of development plans, visions and strategies, and by allowing open access to calculations, by using 'single sourcing', and by sharing ideas on, for example, product development, wherever relevant to cooperation. Information sharing is also likely to result in credible commitments between the involved parties. Commitments can be demonstrated through, for example, long-term contracts, investments in customer-specific assets, employee exchange and joint competency development.

Developing an Implementation Strategy

The development of a formal strategy is a critical step to undertake for every firm implementing JIT logistics. Failure to do so will mean that the significant resources committed (capital, time and people) will be wasted, and the potential benefits of JIT logistics will never be realized. The implementation of JIT logistics is neither a one-time effort nor a short-term project; it is an incremental progression that requires continuous effort. Gélinas (1999) summarized the characteristics of a successful JIT implementation process as:

- an evolution of the organizational culture and a revision of work methods;
- the implementation of new administrative procedures;
- the necessity of not affecting the normal operations of the firm;
- the necessity of respecting the market, and the financial and information technology positions of the firm;
- the acknowledgement of the precedence of some steps over some others.

The implementation of JIT logistics is likely to invoke a change process. Polito and Watson (2006) identified five major constraints upon JIT practices including customer-driven and economic conditions, logistics for continuous supply, organizational culture and conditions, intractable accounting and finance practices, and small supplier difficulties. It is important that firms understand how these constraints would limit the effectiveness of JIT implementation in their organizational and supply chain contexts. A strategy for JIT implementation is not only critical for success, but also provides a framework for defining and prioritizing initiatives for the redesign of logistics activities, system enhancements and organizational restructuring. Such strategy can be derived from a strategic planning process that involves a firm's channel partners. The goal is to align the firm's objectives, as well as those of its partners, with the current status. The planning process should not be confined by the firm's boundaries but represent a careful analysis of the supply chain-wide impact.

The process of developing an implementation strategy involves assessing the internal and external environments. The environmental factors have some bearing on the formation of an implementation strategy for JIT logistics. Internal evaluation relates to the assessment of a firm's strengths and weaknesses, as well as those of its channel members, since no matter how well a firm performs in an area, it will not succeed without the support of its respective partners. Analyzing the external environment involves the evaluation of customers, the business environment, government policies and any other factors that can possibly affect the operations of the supply chain.

Management must bear in mind that implementing JIT logistics is not a short-term project. It is an incremental progression that requires continuous dedication from every employee in the organization. Firms should identify the potential problems of the implementation of JIT logistics and come up with an implementation strategy to overcome the problems. Table 6.1 identifies the common problems associated with the implementation of JIT logistics and the corresponding strategies to tackle the problems.

If JIT logistics is to be effective, then it must be linked with the overall business strategy, effectively aligning the supply chain to fulfil the vision of delivering value to customers. Furthermore, implementation actions should be defined according to a set timetable, with methods of monitoring implementation established. It is necessary for firms to establish strategic control points, which must be monitored in order to achieve organizational objectives. Strategic control points possess five basic characteristics:

1. strategic control points monitor key operations or events;
2. strategic control points are established to identify problems at their point of origin;
3. strategic control points should be capable of indicating the level of performance for many key events;
4. strategic control points should make use of key information in order to be effective;
5. the selection of strategic control points should be balanced. The controls placed on tangible functions, such as production and sales should be balanced with those placed on intangible controls, such as employee development. The two should be balanced such that too strict controls are not placed upon production and too lenient controls placed upon employee development.

Having developed the implementation strategy, the next step is to translate this into a tactical plan, which can be used to traverse the gap between the existing logistics capabilities and those capabilities required to support the business's future direction. This tactical plan identifies and prioritizes the business and system initiatives needed to serve customers more effectively. With the overall strategy in mind, firms later on derive their everyday operational plans.

Table 6.1 Common problems associated with implementing JIT logistics and the strategies to overcome them

Problems in JIT Logistics	Strategies/techniques to overcome the problems
Lack of support	Technical and financial support and incentives from customers, single source/few suppliers base
Lack of top management support	Strategic importance of JIT logistics in productivity and quality, awareness
Low product quality	ISO 9000 certification, TQM, process control
Lack of employee readiness and support	Education, training, motivation by suitable incentive schemes, empowerment
Lack of support from carriers	Supplier is an integral part of purchasing system and long-term commitment
Lack of engineering support	Design for manufacturing, purchasing and application of concurrent engineering
Lack of communication	Computers in JIT purchasing, EDI, EFI, open communications channel.

An Operational Plan for Implementation

An implementation plan sets the tone and direction of the implementation process, while the responsibilities of functional units and their actual day-to-day operations are yet to be formulated. These operational details are to be developed via an operational plan. Considering the possibly extensive scope of the implementation process, what elements should be included in the plan? Operational plans formulate the day-to-day activities and address the specifics of implementation such as who is involved, the allocation of tasks, when activities are to be completed and how the completion of activities are to be realized. Although operational goals are specific to each firm, common goals of JIT include provisions of universal commitment, education and training, facility layout and design, supplier development and management, customer communication, quality improvement, waste identification and reduction, participation, cross-training, job rotation and so on. A study has been conducted to identify the factors that are critical to the success of JIT (Gélinas 1999). These factors are summarized in Table 6.2 and it is therefore desirable to ensure that the operational plan well address these critical areas.

Further to the operational plan, it is important that firms know what the associated costs are, the pitfalls and sometimes subtle costs related to the implementation of JIT

Table 6.2   Operational success factors for the implementation of JIT logistics

Purchasing Management • Partnership with suppliers • Modification of the suppliers network • Suppliers attitude towards JIT • Exclusive and long-term agreements with suppliers • Evaluation, certification and selection of the suppliers • Emergency procedures for late deliveries • JIT transactions • Proximity of suppliers • Deliveries at the point of use • EDI • Carriers adaptation **Inventory Management** • Inventory management compatible with JIT • Information accuracy regarding the stocks • Computerized inventory management system **Quality Management** • Preventive maintenance • Nonconformities prevention • Statistical process control • Control visibility • Total quality • Quality of the supplies • Solving problems at their origin	**Production Management** • JIT production • Automatization and computer technologies • Compliance to the fabrication programme • Tuning, set-up and repair time minimization • Information accuracy • MRP • Kanban or another production control system that suits JIT needs • Time buffers planning • Production stops in case of problems **Localization and Layout** • Proximity of work stations • Material path management • Flexible production cells **Distribution Management** • Carriers adaptation • Distribution network configuration • Single or multi-echelon distribution • Warehousing facilities

logistics. Some of the costs are due to implementation issues, which may require a firm to redesign its logistics processes to better fit the implementation requirements of JIT logistics. Other costs are linked to inventory flows through the system, as well as the costs related to procurement and communication to suppliers. It is also possible for unexpected events to occur that can be disruptive to the logistics processes that run on lean inventories. To tackle these operational issues, managers in a JIT environment must use CI to search for ways to make the implementation more effective. The CI requirement of JIT is an important area that firms should emphasize. As part of the implementation of JIT logistics, firms must instil the habit of expecting continuous small improvements in the process. Employees must never be satisfied with the current environment, but always be moving closer to the ideal situation for cost and service improvements.

One important operational issue for the implementation of JIT logistics is how to incorporate it in the overall business strategy. This stage also includes making clear which resources shall be allocated, and who will be involved in the implementation process and so on. The implementation of JIT logistics internally and in relation to cooperative partners is a process that affects logistics activities in many areas. These areas include purchasing, production planning and distribution. It is therefore crucial that top-level managers follow the process with interest and are active and prepared to maintain the direction, and are also capable of executing the necessary decisions.

Another important operational issue is how to build team efforts in the implementation process. A team is more than just a group of people involved in a common project. A team is made up of individuals with complementary competencies, who are enthusiastic about a shared goal for JIT logistics. This goal includes specific and operational performance targets for which the team players hold each other responsible. A dividing line can be drawn between intra-organizational teams and inter-organizational teams. Teams of employees from sales/marketing, logistics, production and accounting exemplify intra-organizational teams. This type of team is necessary in order to establish a process thinking in a firm in order to coordinate activities across line functions. Inter-organizational teams are composed of representatives from different partner firms in the supply chain. These teams are important connection points between the supply chain participants.

Teams and other organizational forms that are designed to promote cooperation across functions and companies should be supplemented with incentive programmes that reward the total efforts of the team. If wages and career possibilities continue to be based on individual performance and competition, then employees cannot be expected to put themselves wholeheartedly into teamwork. It is also important that team goals for the implementation of JIT logistics are well defined, operational and directly related to the firm's overall goals and vision. If this is not the case, then good results cannot be anticipated from the team's efforts. Correspondingly, it is important that a firm's overall vision and goals are articulated clearly, so that the challenges of fulfilling the articulated goals are clear to each team member.

It is also important to ensure fair sharing of the advantages and risks with the implementation of JIT logistics, making individual participants feel that the rewards of entering into the cooperative effort are evenly balanced between the resources invested and the risk of loss. Otherwise, the incentive to continue cooperating disappears. At the beginning of implementation, there is a harmonization of the expectations and goals between the actors in the logistics activities. The resource contribution will often be unevenly divided among the cooperating partners. Therefore, it is important with regard to long-term cooperation that participants who invest a relatively large amount of resources receive a corresponding proportion of the rationalization gains. This equitable distribution can help ensure long-term contracts, joint investments in production equipment and agreements as to the sharing of cost savings.

Pilot Projects

Pilot projects for the implementation of JIT logistics offer firms a relatively low-cost means to experiment and test the feasibility of new ideas. Pilot projects involve the actual implementation of the new ideas on a smaller scale, for example, a product line, a specific delivery route and a number of specific branches of the intermediaries and so on. The goal is to realize the JIT principles in the tested subjects so as to ensure that the operational plans are feasible, efficient and effective.

It is a desirable choice for firms to implement a number of small, relatively risk-free pilot projects, to test their ideas prior to a broader roll-out if successful. The projects that do fail will be taken as learning experiences, so that similar mistakes are not repeated. Another reason for the pilot projects is that in today's highly competitive environment logistics activities that do not function properly will not survive for long. Pilot projects help in the fine tuning of logistics solutions before they are rolled out. New processes are perfect candidates for pilot projects, so as to avoid damaging customer relationships while new processes are implemented on a broad scale. The pilot projects for the implementation of JIT logistics should include the following elements:

- involvement of key stakeholders, suppliers, customers and employees;
- selection of scope and environment: which location, business, group of items or customer should be the host the pilot? This activity should focus avoiding risk whilst ensuring exposure to a wide range of business scenarios;
- identification of the key questions that the pilot must answer: what are the key success factors?

It becomes critical to measure the results of pilot projects carefully. This should ensure that the expected benefits are achieved and any adjustments are made before rolling out the project more broadly across the firm. While pilot projects for the implementation of JIT logistics offer benefits to the firm, they do possess some limiting characteristics. The advantages include the following:

- The pilot project allows problems that have not previously been considered to surface. Employees and managers can work through these problems and learn from their experience. Lessons learned can be applied to other implementation efforts.
- The operational plan for implementation can be applied and its effectiveness evaluated prior to full-scale implementation. The pilot project provides the opportunity to revise the strategy and plan in the light of problems that were not considered previously.
- Pilot projects gradually ease the firm into implementation. The result of this small-scale to full-scale implementation is to allow employees to become accustomed to the idea of change.

The most common disadvantage of pilot projects includes limiting the benefits received to a small area of the firm and the experience with the project to a few employees. Overcoming this limitation is possible through establishing several pilot projects in parallel. Establishing several projects that run independently of one another also increases the firm's chances of success. Failure or excessive problems associated with one pilot project have no bearing on the success of the other projects and the firm is less likely to give up future implementation efforts.

Dealing with Employee Resistance

JIT regards change as inevitable and essential to the growth of a firm. Unfortunately, most people within a firm despise and resist change. They are afraid of change and resist it by nature, because it affects their established ideas and opinions. Disappointments from previous experiences with activities involving change may account for this. Especially in the initial phase of the implementation of JIT logistics, resistance seems to be very high because certain people within the firm see this as a threat due to the related uncertainties and obscurity. Other common reasons for people to resist are: the need for additional information, deep mistrust and lack of understanding. Because of this, conflicts emerge whereby people frequently refer to the old reliable and familiar situations.

Change is not something that any department and individual takes up easily, and administering changes in organizational practices has to be considered with care. Senior managers should bear in mind that some people, although a minority, will be 100 per cent negative to the concept and principles of JIT. These people should be identified and worked upon to encourage a more positive perception. Senior management should not underestimate the impact they can have on these people. However, if these people are not prepared to change their behaviours and attitudes, harsh decisions will be required. Methods to handle resistance are:

- Top management must communicate face-to-face and give information about the what, why and how of change.
- Be honest about the actual situation. State clearly how long the change will last and what the consequences will be on the quality of employees. Provide information on time. Silence creates doubt and usually causes rumours to spread, which undermines their trust in management. Don't provide too much information at once, because employees need time to absorb the information.
- Support the proposals with clear arguments.
- Inform the employees about the advantages of change and how the gap between future and present situations must be closed.
- Have meetings with those who show clear resistance and give a detailed reaction to all their objections.
- Involve concerned employees in the project.
- Involve key persons and the union in the decision-making process. After all, if the stakeholders are involved in making decisions, chances of acceptance will be much higher and therefore, also the effect. Involve also change agents to facilitate, monitor and encourage change.

- Put the project on hold if there is too much resistance and you are not able to count on the support of the majority.

As discussed earlier, employee involvement represents a key element to the success of JIT. The implementation of JIT logistics is a change process and resistance to change is likely to be encountered. The most relevant and significant fundamental cause of implementation difficulties rests with employee resistance to the JIT change. Managers must realize that employees are the most valuable organizational assets, for it is through them the goals and continuation of the firm will be realized ultimately. The earliest stages of implementation should mark the beginning of a new relationship with employees, characterized by trust, equality and respect.

Setting examples through management actions and consistent behaviour primarily serve to develop trust in the management/employee relationship and set standards that clearly apply to all within the firm. Employees will gradually come to realize that management is committed and prepared to make changes that affect them. Through consistent behaviours, management sets implicit standards or norms and enables employees to recognize that this is how it is going to be. Employees can then establish realistic expectations of JIT logistics, and assess their own behaviours and attitudes.

There will often be some departments that react by withholding information and working against internal cooperation. There can be many reasons for this type of reaction. One way of altering this behaviour can include creating goals and reward systems. These systems can motivate cooperation and interdepartmental information exchange. Management should follow a proactive approach to communicating with the employees. Examples of a proactive approach include involving employees, managers setting examples through 'doing as they say', consistency of behaviours and providing employees with education and training.

Involving employees in the implementation effort must extend beyond simply obtaining employee input for the change, although this in itself cannot be disregarded. Rather, involving employees must include seeking employee approval and providing them with access to information previously not shared. Information regarding available resources and progress towards goal achievement is necessary to keep employees informed of the activities of the business. Sharing with employees to this extent will present the greatest difficulties for management, especially if the previous working environment was characterized with 'us versus them' attitudes and closed-door discussions.

Adopting Enabling Technologies

Adoption of enabling technologies in a collective manner in a supply chain can benefit all the participating firms by facilitating information sharing. It is essential to understand the factors affecting technological adoption at firms and the performance implications, particularly for issues relating to: 1) why enabling technology is (or is not) adopted; 2) what factors affect the intensity (level) at which enabling technology is adopted in organizations; and 3) the performance implications of the extent of technological adoption for logistics.

Rogers (1983) defined innovation as an idea, practice or object that is perceived as new by an individual or by other units of adoption. This suggests that an innovation need not be a matter of newly invented products/processes, but rather something new to the organization adopting and using it. The implication is that technological adoption is not just concerned with the technology itself, but also with the process of adoption. Thus, the conditions and factors that affect technological adoption play a significant role in attaining the desired results of the adoption of a technology.

Generally, there are five factors that are critical to determining the level at which enabling technologies are adopted in organizations (Lai et al. 2006b). The factors are: 1) technological opportunism; 2) organizational innovativeness; 3) complementary assets; 4) perceived advantages; and 5) institutional pressures. How and why these factors affect the adoption of enabling technologies in organizations and their potential effects are elaborated in the following.

TECHNOLOGICAL OPPORTUNISM

Technological opportunism is the sense-and-response capability of organizations with respect to new technologies. Technology-sensing capability refers to the ability of an organization to acquire information on new technological knowledge and innovations. It is a competence in sensing, identifying and evaluating new opportunities and potential threats to an organization. Technology-response capability is an organization's ability and willingness to respond to the new technologies it identifies in its environment that may influence the organization (Srinivasan et al. 2002). It represents the firm's ability to exploit the opportunities and to mitigate the threats posed by the technology. When a technologically opportunistic organization is aware of technological changes in its environment, it is presumed that the organization is likely to adopt and use the technology and takes this investment as a potential source of growth. As a result of perceiving the adoption of technology as a strategic act, a technologically opportunistic firm proactively adopts new technology and motivates its business units to utilize the technology to seek the greatest strategic benefits from the adoption. For example, when Wal-Mart demands its suppliers to implement RFID technology, retail analysts at Sanford C. Bernstein estimated that the firm can save as much as US$8 billion annually (Boyle 2003). Other retailers such as Target and Albertson's followed the trend of adopting RFID and started to test RFID applications in their supply chain activities. Following this line of reasoning, it can be speculated that firms that are technologically opportunistic tend to seek and adopt new technology, and that the extent of technological adoption in the context of SCM is a manifestation of this postulation.

ORGANIZATIONAL INNOVATIVENESS

Chandy and Tellis (1998) suggested that firms must be able to overcome their inertia and innovate to sustain their business. An increasing number of studies support the notion that innovativeness is the key to achieving superior organizational performance. The innovativeness of an organization is defined as the 'generation, acceptance and implementation of new ideas, processes, products and services' (Thompson, 1965). Zaltman et al. (1973) argued that innovativeness is an organizational characteristic that refers to openness to new ideas as an aspect of a firm's culture, which is a measure of the organization's orientation towards innovation. In the context of SCM, organizational innovativeness reflects an organization's willingness and ability to use technologies to enhance collaboration among partner firms in the supply chain. For example, the suppliers of Wal-Mart should overcome their inertia on EPC adoption arising from their organizational characteristics such as organizational structure and management practices. They should seek to reap the benefits of adopting RFID, not just to satisfy the requirements of Wal-Mart, but also to better utilize the technology to improve their logistics operations.

COMPLEMENTARY ASSETS

Complementary assets are those that help an organization to obtain value from technologies and positively affect the process of adopting technology (Rogers 1983). Complementary assets sustain the competence of an organization. The ownership of complementary assets, particularly specialized and/or co-specialized assets, determines which organizations win or lose from a technology (Teece 1986). For example, Paxko is a global manufacturer of paper, plastic and aluminium consumer products and food service packaging. It has 12 manufacturing plants in North America, six in Europe, two in South America and one in China. For systematic control and coordination of operations within and between its manufacturing plants, EPC adoption enables Paxko to improve its production, asset utilization, inventory reduction and labour productivity, for example, machineries and equipment are tagged to track their usage, location (within or borrowed by another plant) and maintenance (Chappell et al. 2002). The IT-enabled systematic coordination of operations within and between plants serves as a complementary asset for firms to collect detailed, accurate and timely information for performance enhancement. Prior studies have suggested that when incumbent firms possess specialized complementary assets that retain their value despite shifts in technology, these assets protect the incumbents from the effects of the destruction of competence (Tripsas 1997). The existence of a pre-existing base of knowledge or the capacity of an organization to use a technology can lower the costs of learning the technology in the organization (Cohen and Levinthal 1990).

PERCEIVED ADVANTAGE

Relative advantage is the degree to which a technology is perceived as better than the idea it supersedes (Rogers 1983). It indicates the strength of the reward or punishment, which is measured in economic terms, and of the factors of social prestige, convenience and satisfaction resulting from the implementation of a technology. An organization is likely to implement a technology if it perceives that the technology offers relative advantages over its status quo (Tornatzky and Fleischer 1990). For example, perceiving the advantages of adopting RFID to improve its supply chain operations, Wal-Mart mandates its top suppliers to implement RFID technology in an attempt to create an efficient and compatible base of trading partners in its supply chain. Given the traits of EPC that promote efficient and effective trade and product tracking among firms in a supply chain, organizations interested in achieving integration in their supply chains and enhancing their capability are likely to adopt and broadly use EPC.

INSTITUTIONAL PRESSURES

King et al. (1994) concluded in their study on the impact and regulatory role played by institutions involved in information technology innovation that institutional influences and regulations can be construed as the ideologies governing supply-push and demand-pull models, which are consistent with the scenarios prevalent in supply chains. At the inter-organizational level, institutional pressures from government, industry alliances and social beliefs define socially acceptable conduct. The social pressures common to all firms in the same sector lead firms to exhibit similar structures and activities (DiMaggio and Powell 1983). The adoption of RFID in the retail industry shows how institutional pressures affect the adoption of information technology (Lai et al. 2006a). For example, a supplier that relies heavily on big retailers for businesses, such as Wal-Mart and Tesco, is likely to comply with the requests of their dominant customers on the adoption of RFID technology to retain their businesses. Organizational sociologists have long argued that firms adopt technologies because of institutional pressures from constituencies in their environments. As a technology spreads, a threshold is reached beyond which implementation is motivated by legitimacy rather than by the belief that it will improve performance. Thus, in the context of SCM, the adoption of an appropriate information technology is a form of organizational response to institutional pressures (Lai et al. 2006b).

Diffusion of JIT logistics to Partner Firms

Before embarking on any JIT initiatives for logistics, it is important to consider the diffusion, market acceptance and legitimacy of JIT logistics in the supply chain to facilitate and support communication among supply chain partners to ensure its effective implementation.

Notwithstanding the intuitive value of JIT logistics, without its acceptance and diffusion in the supply chain, it is unlikely that its implementation will fully realize its operational and economic benefits. The widespread adoption of JIT logistics is necessary if its full operational and economical benefits are to be realized. This is because the network effects will increase the payoffs to adopters as they can share the cost and service benefits via JIT practices in their logistics activities. The diffusion process can be accounted for by reference to institutional isomorphic influences on the adoption of JIT principles in the realm of logistics.

Diffusion is the process by which an innovation is communicated through certain channels over time among the members of a social system. A major pitfall of the concept of diffusion is it assumes that organizations within a group are free and independent to choose to adopt (or not to adopt) an innovation, for example, JIT logistics. It fails to address the institutional isomorphic processes, which can affect the decisions of organizations on the adoption of JIT logistics. Supplementing the concept of diffusion, institutional isomorphism, which refers to 'the constraining process that forces one unit in a population to resemble other units that face the same set of environmental conditions', provides a theoretically sound basis to explain the adoption of JIT logistics.

According to DiMaggio and Powell (1983), institutional isomorphism considers 'the major factors that an organization must take into account are other organizations'. In addition to competing for resources and customers, firms are competing for political power and institutional legitimacy for social and economic rewards. The implications of the institutional forces are that firms may base their decisions on the adoption of JIT logistics on one or more of the following mechanisms: 1) they may experience pressure from other firms upon which they are dependent; 2) they may mimic other firms within their sector that they perceive to be successful; and 3) the professional associations may exert pressure on the firms by establishing a cognitive base and legitimization for the autonomy of the industry.

To diffuse JIT logistics practices to other firms, it is desirable to understand the influence of the different types of institutional isomorphic processes from the views of both the initiators (that is, JIT-adopted firms) and the followers (that is, JIT-adopting firms), where each process has its own objectives, attributes and injunctions or compliances in the adoption of JIT logistics in the supply chain. In the following, we discuss the three institutional isomorphic processes, namely coercion, mimesis and norms, for the adoption of JIT logistics in the supply chain. The discussion will shed light on the diffusion of JIT in the realm of business logistics, which will help managers to better understand and address the issues of institutional isomorphic influences that affect the adoption of JIT logistics in the supply chain.

Coercion

The coercive pressures are exerted on a dependent firm by other firms and by cultural expectations in the society within which the dependent firm operates. Firms that have adopted JIT logistics are likely to ask their trading partners to do the same to ensure consistency in logistics practices to maintain the efficiency of their supply chains. Dependent firms that rely heavily on a dominant firm's business for survival will comply with their dominant partner's direct imposition of the requirement to adopt JIT logistics. In this way, the dependent firms are coerced to adopt JIT logistics for close collaboration and efficient operations in order to join forces or continue to participate in the supply chain of the dominant firm. In other words, the dependent firms are willing to adopt JIT logistics in order to demonstrate their commitment to the trading relationship, thus displaying their conformance to legitimacy.

The adopting firms may acquire technical assistance from the dominant firm and/ or relevant professional associations, especially in the case of integrating their existing systems and processes with the newly adopted JIT logistics practices. In some cases, the initiators will instruct their partners to adopt JIT logistics without providing assistance. The followers will be compelled to acquire the necessary information and help from relevant professional associations if they wish to participate in the supply chain. On the other hand, in some instances, the dominant firm provides technical support and shares its experience with the adopting firms to ensure the quality and conformity of the adoption of JIT logistics.

In the case of the coercive process, initiators face the risk of losing their investment in assisting followers. For instance, followers may not be able to implement JIT logistics in their processes to realize the benefits themselves and may decide to withdraw from the supply chain relationship. Also, initiators have to contend with high switching costs (for example, technical help and training costs) in the supply chain relationship by offering support and assistance to followers. On the other hand, although the dominant firm provides support and assistance for the adoption of JIT logistics, the adopting firm bears the risks of revealing its internal processes and trade secrets to the dominant firm. As a result, the adopting firm may end up with higher operating costs if the dominant firm shifts activities and costs (for example, products storage and handling) to its followers.

Mimesis

Another type of institutional isomorphism is the force of uncertainty that encourages the imitation of practices. When a firm has ambiguous goals and operates in a volatile environment, it models itself on other organizations, especially on the organizations that are closely associated with it, in response to the uncertain business environment. The followers may not be aware of their mimetic behavior as the firm being imitated may merely serve as a convenient source of practices.

Discovery and learning of JIT logistics may occur indirectly through industry associations, employee transfers, employee turnover and information interchanges. When firms face a chaotic and uncertain environment, they try to outperform their competitors through low cost or differentiation. In the context of business logistics, when seeking strategic logistics tools and practices to outperform competitors, firms consciously or unconsciously mimic the practices of their supply chain partners because:

- the firms have easy access to the logistics practices of their partner firms through information interchanges and inter-firm process integrations;
- the attributes that seem to account for the successful logistics practices of the partner firms are easily observed by the mimicking firms in the supply chain;
- the partner firms are often willing to share logistics experiences and know-how with one another, as the sharing of information and knowledge benefits them as well while they serve the same supply chain.

The convenient access to the logistics practices of partner firms, the recognition of critical success factors and the dissemination of logistics know-how and experience all lead to the intentional or unintentional imitation of practices ranging from diffusion to industry-wide application of JIT logistics in the supply chain. Firms tend to model themselves on similar organizations in their field that they perceive to be legitimate or successful, by imitating and acquiring the attributes, innovations and practices that have been proven to be attributable to the success of the supply chain.

To encourage the voluntary adoption of JIT logistics, firms that have adopted JIT logistics create awareness of JIT logistics in their supply chain by demonstrating improvements in their intra- and inter-organizational processes via JIT logistics. Also, by sharing their experiences of the processes of the adoption of JIT logistics, the initiators alleviate the doubts of their followers about adopting JIT logistics and reduce their perceived risks associated with the adoption. However, in the case of mimetic institutional isomorphism, the initiators tend to avoid providing technical support to JIT logistics-adopting firms by only creating awareness and sharing knowledge of JIT logistics. The initiators may miss the opportunity to build partnerships with their suppliers and customers, who might face difficulty in justifying the associated investment in order to realize the benefits of the adoption of JIT logistics. Furthermore, the firms that are mimicking the adoption of JIT logistics may refrain from conforming to the legitimate practices of the supply chain and from committing to the supply chain. The adoption of JIT logistics by the followers is mainly for the purpose of improving their supply chain performance. In other words, business commitment from the initiators is not necessarily generated in the processes of the followers in imitating the adoption of JIT logistics for improving their supply chain.

Norms

Normative processes are the third type of institutional isomorphism. They stem from professionalization, which is concerned with the establishment of legitimization for the autonomy of a supply chain. Firms in a supply chain are subject to the norms, standards and expectations of their logistics practices in order to attain effective coordination. Generally, firms that have adopted JIT logistics are unlikely to establish new partnerships with non JIT logistics-adopters, as this would require the JIT logistics-adopted firms in the supply chain to set up additional procedures to handle the logistics processes. This, in turn, would have a potentially negative impact on the efficiency of the supply chain. In other words, a supply chain operates under the normative institutional isomorphic process, whereby firms that are to become eligible participants in the supply chain are expected to adopt JIT logistics. While normative and coercive institutional isomorphic processes are similar in nature, that is, enforcing standard logistics practices in a supply chain, they are indeed different in practice. With normative institutional isomorphic process, the followers adopt JIT logistics voluntarily, even though there is no commitment for business; while firms mandated to adopt JIT logistics via the coercive institutional isomorphic process are guaranteed to receive business from their supply chain partners.

In the normative institutional isomorphic process of adoption, the initiators are less likely to provide support and share experiences with the potential JIT logistics-adopting firms. This is because the initiators have already formed an efficient and effective process of supply chain coordination by adopting JIT logistics, which guarantees the autonomy of the supply chain. Hence, no business commitment or obligation is created or maintained in the supply chain, as the participants have the option of deciding whether to adopt JIT logistics with the intention of gaining business opportunities. Table 6.3 summarizes the dimensions of institutional isomorphism on the adoption of JIT logistics in the supply chain.

The discussion above highlights the view that the adoption of JIT logistics in the supply chain is subject to the influence of three types of institutional isomorphic process, that is coercion, mimesis and norms. The different types of institutional isomorphism on the adoption of JIT logistics should be recognized. It is therefore important to consider these institutional isomorphic processes in the diffusion of JIT logistics in the supply chain.

Data Collection and Measurement Systems

To evaluate and improve the progress of the implementation of JIT logistics, performance measurement is necessary. Data collection and measurement systems supplement the operational plan for JIT logistics as they provide the mechanisms to collect all the information

Table 6.3 Dimensions of institutional isomorphism on the adoption of JIT logistics in the supply chain

Initiators (Adopted Firms)	**Objective**	Striving for supply chain performance improvement	Seeking supply chain performance improvement	Striving for supply chain autonomy
	Attributes	May provide technical support, and share knowledge, know-how, and experience to followers Offering business commitment	Creating awareness Sharing know-how and experience Providing minimal or no support to followers Do not offer business commitment	Standardizing practices in supply chain Providing minimal or no support to followers Do not offer business commitment
	Injunction	Forcing JIT logistics adoption with or without assistance and support	Fostering the voluntary adoption of JIT logistics	Imposing the adoption of JIT logistics
	Possible Drawbacks	Incurring switching costs	Losing business opportunities	Losing business opportunities
Followers (Adopting Firms)	**Objective**	Attaining business commitment	Seeking supply chain performance improvement	Seeking business opportunities
	Attributes	Acquiring knowledge and experience in JIT logistics Seeking help for technical support Conforming to legitimacy	Responding to uncertainty Voluntarily seeking information about JIT logistics	Seeking entry to a supply chain
	Compliance	Adopting due to coercion	Intentional or unintentional imitation.	Adopting due to norms
	Possible Drawbacks	Revealing internal processes to initiators Increasing the cost of operations due to shifting activities and costs from initiators	No business commitment guaranteed by supply chain partners	No business commitment guaranteed by supply chain partners

on a continual basis, as well as providing the means with which to compare actual performance to established standards. Data collections systems must provide information that is timely, relevant, consistent and complete. Employees should receive training to utilize the systems properly, as their effectiveness is as much a function of the systems themselves as it is the users. Common data collection systems include databases, where all the information is fed into and retrieved from the database when needed by use of a computer, or manual standardized procedures where data are gathered and located in designated areas.

Measurement systems for JIT logistics serve two purposes, namely providing the basis for monitoring implementation activities, monitoring or tracking the progress of work, and providing a standard for improvement activities. Although a firm is likely to have measurement systems in place prior to the implementation of JIT logistics, these traditional measurements are not likely to reflect JIT organizational requirements.

Measurement systems for JIT logistics differ from traditional measurement systems on three levels, namely they must relate to organizational strategic objectives, provide information that reflects the organizational direction and relate the firm's performance in terms of improvement.

Measurement systems for JIT logistics must be accurate, effective and meaningful. The measurement must be consistent with a 'total business concept' or the view that a firm is operating as a system or integrated whole. Total business concept measurements provide information related to performance at the cell level, as well as for the firm as a whole. Measurement systems for JIT logistics based upon the total business concept will provide a firm with the following:

- accurate information – measurement systems that are capable of providing information free of errors will save the firm valuable time and effort in the long run;
- information about the waste present in the logistics activities;
- information that relates actual performance to pre-established plans;
- awareness of the timeliness of information from logistics activities.

The actual performance measurements for JIT logistics can be grouped into several categories including those that measure inventory, delivery performance, quality, production and data accuracy/paperwork reduction measurements. Measurements such as these provide examples of effective measurements; however, their application within a firm may vary, depending upon the nature of the firm's product or service line. Table 6.4 compares the performance measurements of implementing JIT logistics between early stages and advanced stages.

Performance Measurement

There is a saying that: 'If you cannot measure it, you cannot control it. If you cannot control it, you cannot manage it. If you cannot manage it, you cannot improve it.' The need for performance measurement for JIT logistics is obvious. Performance measurement evaluates how the output of a process conforms to requirements and expectations, and suggests how well an individual, process or team is operating. The criteria most commonly associated with performance in logistics include costs (or efficiency), service (or effectiveness in meeting customer requirements) and availability or schedule (or cycle

Table 6.4 Performance measurements for JIT logistics in early to advanced stages

	Early stages of JIT logistics	**Advanced stages of JIT Logistics**
Information systems	<ul><li>MRP II systems in production planning</li><li>No EDI links to external cooperation partners</li><li>Low/ poor degree of system integration</li><li>Bar coding only on final products</li><li>Internet/Extranet used primarily for correspondence</li></ul>	<ul><li>ERP system implemented</li><li>APS systems used as decision support</li><li>EDI links to important cooperation partners</li><li>Use of bar codes to track-and-trace throughout the supply chain</li><li>ECR with important customers</li><li>Internet/Extranet used in purchasing and sales</li><li>VMI with selected customers</li><li>CRM/supplier relationship management for management of customer/vendor base</li></ul>
Organization form	<ul><li>Primarily functional oriented</li><li>Logistics/SCM not represented at the level of the director</li><li>Fragmented logistics function</li><li>SCM not a part of the company's business model</li><li>Few points of contact between the companies in the supply chain</li></ul>	<ul><li>Primarily process oriented</li><li>Logistics/SCM represented at the director level</li><li>Teams, both across functional borders and between companies</li><li>SCM is an important component in the company's business model</li><li>Many contact points between companies</li><li>Focus on management of relations regarding the use of supplier relationship management and CRM</li></ul>
Information exchange	<ul><li>Ordering by fax, phone, or e-mail</li><li>Access to customer/supplier warehouse status</li><li>Harmonizing of warehouse stocks in the supply chain</li></ul>	<ul><li>Ordering by Internet/Extranet</li><li>Production plans and sales prognoses accessible for suppliers</li><li>Vendors included in product development</li><li>Mutual access to cost calculation</li></ul>
Time dimension	<ul><li>Contracts of limited duration</li><li>Regular bidding rounds to test the market</li></ul>	<ul><li>Contracts with longer timelines and agreed-upon efficiency goals</li><li>Continuous benchmarking in order to secure high efficiency and quality</li><li>Contract negotiation without market testing</li></ul>
Measuring and management	<ul><li>Measuring of delivery service</li><li>Measuring of the degree of warehouse service</li><li>Measuring of logistics costs</li></ul>	<ul><li>Performance goals on all business processes</li><li>Continuous vendor evaluation</li><li>Measurement of customer service</li></ul>

time). The best practice is to set targets for performance and then measure conformance against the set targets. These measurements provide a quantitative way of judging performance and can serve as a basis for learning how to make improvements.

A performance measurement and reward system should align with the JIT philosophy as it is necessary for management to realize the benefits derived from implementing JIT logistics practices and be aware of how well the employees are maintaining the JIT standards. Traditional accounting criteria may not well reflect the values of JIT at once. A management accounting system that includes non-traditional indicators such as product quality, on-time delivery, set-up times and inventory turns will be able to reveal a firm's JIT performance.

During the past decade, many firms have worked to improve their processes and performance by developing a balanced scorecard approach to self-measurement. This highlights the importance of tracking performance in a number of different areas, rather than relying purely on sales figures, revenue results or other purely bottom-line related measures. In the early 1990's, Kaplan and Norton introduced the concept of balanced scorecard performance measurement. They argued that firms and managers are not effectively capturing their true performance when they track only financial measures. While financial measures are unambiguous, they do not show other less-quantifiable but still critical measures of performance. Firms need to assess their performance by looking forwards as well as backwards.

Kaplan and Norton (1992) pointed out that a balanced scorecard could be used to not only measure past performance, but also model future strategy and build a cohesive approach for future competitive success. The scorecard becomes a strategic management system for achieving long-term goals, allowing organization leaders to set future performance standards even while guiding current activity. Accordingly, they suggested that firms create a unique scorecard reflective of their own businesses that contain four key perspectives:

1. Financial objective – how do we look at our shareholders?
2. Customer outcomes – how do our customers perceive us?
3. Internal business processes – in what areas must we excel?
4. Learning and growth – how can we continue to grow and develop?

This balanced scorecard approach is helpful for firms to understand their performance drivers, articulate their business strategy and communicate that strategy effectively, and align initiatives towards the common goal of JIT logistics. The goal is to deliver superior customer value at the lowest cost possible by managing the upstream and downstream relationships with suppliers and customers. The question is how can a firm assess how well its logistics activities are being managed? The challenge of evaluating logistics performance lies in the fact that logistics activities encompass processes beyond organizational boundaries and therefore make the traditional performance measurement that solely relies on internal processes inadequate. In addition, how can a firm ensure that all its stakeholders, for example, suppliers, customers and the firm itself are satisfied? Different parties may have different expectations towards the logistics activities.

From customers' point of view, they usually do not care too much how about how the product is transferred but are concerned with whether the product is delivered in good shape and on time, and therefore reliable delivery service. On the other hand, a supplier would like to ensure that the products being transferred are always maintained in good condition since damaged items incur a cost (as handling damaged items does involve extra effort in addition to the product replacement costs). A supplier will also demand the items to be delivered on time since it directly affects customers' perception of its service quality. Alternatively, when a firm receives a customer's order, it will always seek the most cost-effective approach, for example, delaying shipment until carriage in full truck loads, to serve the request. Whether the products are transferred in good shape is also a concern since the firm will have to pay for the damaged items. It is obvious that the expectations of stakeholder groups differ. The demand of one group is probably at the expense of at the other. For instance, premium delivery service for customers leads to high delivery costs for the firm. This is due to the fact that the output of a member firm is indeed the input of another member in the logistics process. The objective of logistics is to satisfy customers, both upstream and downstream, with higher effectiveness and efficiency than the competitors. Therefore, performance measurement needs to incorporate, as well as balance, both the operation efficiency parameters and service effectiveness measures (Lai et al. 2002).

To better measure the performance of JIT logistics, the SCOR model, developed by the Supply Chain Council (www.supply-chain.org), provides a useful performance measurement evaluation framework. The SCOR model views activities in the supply chain as a series of interlocking inter-organizational processes with each individual firm comprising five components, namely plan, source, make, deliver and return, which are described as follows:

- Plan – the tasks of planning demand and supply set within an overall planning system that covers activities such as long-term capacity and resource planning. Examples of performance measures include costs of planning activities, inventory financing costs, inventory days of supply on hand and forecast accuracy.
- Source – the task of material acquisition, set within an overall sourcing system that includes activities such as supplier certification and supplier contracting. Examples of performance measures include materials acquisition costs, cycle times for receiving and using goods, and raw materials' day of supply on hand.
- Make – the task of production execution, set within an overall production system that include activities such as shop scheduling. Examples of performance measures include production cycle time, number of product defects and other quality issues.
- Deliver – the day-to-day tasks of managing demand, orders, warehouse and transportation, and installation and commissioning. These tasks are set within an overall delivery management system that includes order rules and management of delivery quantities. Examples of performance measures include fill rates, order management costs, order lead times and transportation costs.
- Return – the return of nonconforming goods for replacement or rectification, and the recycling of materials no longer needed by the customer. Examples of performance measures include number of complaints, speed of customer service calls and customer follow-up measures.

Each of these components is considered a critical intra-organizational process in the supply chain with four measurement criteria:

1. supply chain reliability
2. responsiveness/flexibility
3. costs
4. assets.

The first two criteria deal with effectiveness-related (customer-facing) performance measures, whereas the other two are efficiency-related (internal-facing) performance measures of a firm. Customer-facing measures relate to how well a supply chain delivers products to customers. Internal-facing measures are concerned with the efficiency with which a supply chain operates. The SCOR model provides useful indicators on how effective a firm is using its resources to create customer value. It considers the performance expectations for member firms on both the input and output sides of supply chain activities. The performance measures of the SCOR model are summarized in Table 6.5.

The key question of JIT logistics is how to coordinate the efforts of every firm in the supply chain and every employee within those firms. The coordination must provide ever-increasing amounts of value added to customers willing to pay for it. Performance measures drive behaviour in any system. The selection of performance measures is crucial inside a firm and throughout the supply chain. Managers coordinate the behaviours of their employees and of their partners in the supply chain by the use of performance measures. It is through the use of measures that firms can determine if they are making progress towards the goals of JIT logistics.

Table 6.5 SCOR performance measures

Supply chain process	Measurement criteria	Performance indicators
Customer facing	Supply chain reliability	Delivery performance Order fulfilment performance Perfect order fulfilment
	Flexibility and Responsiveness	Supply chain response time Production flexibility
Internal facing	Costs	Total logistics management costs Value-added productivity Return processing cost
	Assets	Cash-to-cash cycle time Inventory days of supply Asset turns

A Ten-step Approach for the Implementation of JIT Logistics

This generic ten-step approach provides a roadmap for carrying out a systematic, gradual and team-wise method for the implementation of JIT logistics. The approach offers procedural guidelines for firms contemplating the implementation. The ten steps are displayed in Figure 6.1 and elaborated below.

STEP ONE – MANAGEMENT COMMITMENT

The first step in implementing JIT logistics is to solicit and capture the commitment and support of management so as to make the implementation efforts sustainable. The lack of senior management commitment, awareness and vision in an organization is often diagnosed as the main cause for the failure of implementing JIT logistics. The importance

160 Just-in-Time Logistics

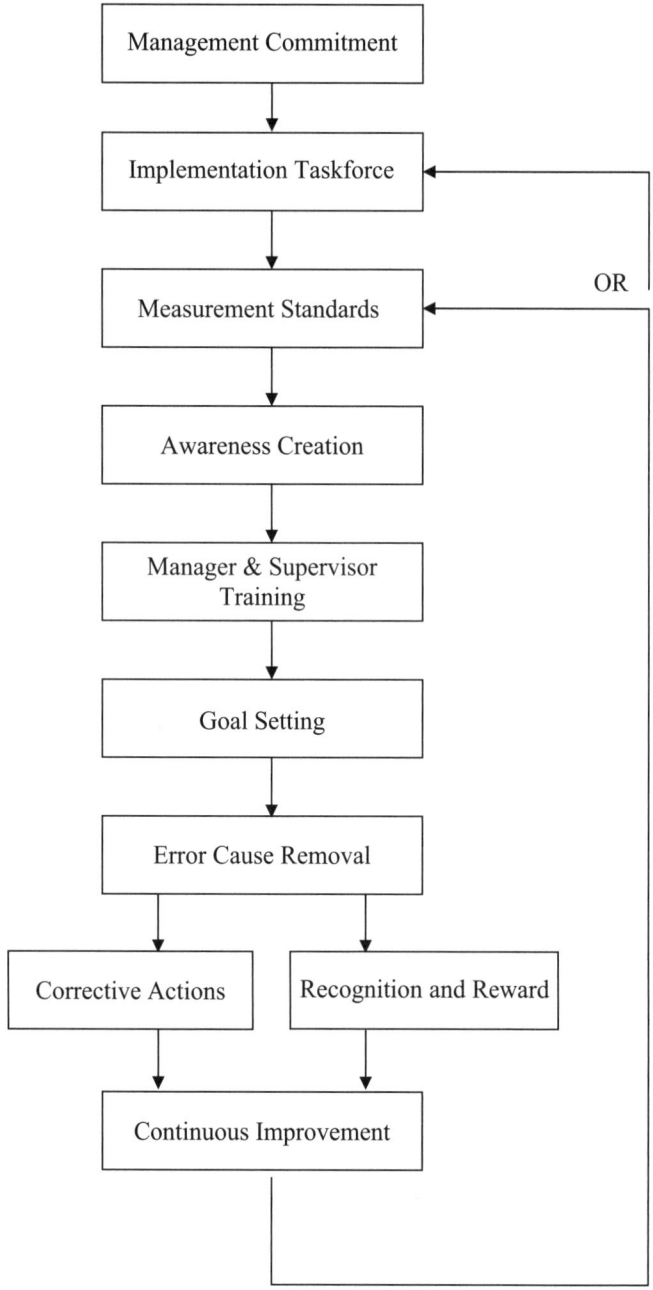

Figure 6.1   A ten-step approach for the implementation of JIT logistics

of top management in providing a clear and strong message about its vision for JIT logistics cannot be over-emphasized and must be articulated at all levels of the firm. The purpose of this step is to make it clear where the top management stands on JIT logistics, pronouncing it throughout the firm. To start with, the need to implement JIT logistics is extensively debated and discussed by the top management team, including the chairman and the CEO of the firm. The vision for the implementation of JIT logistics is defined, which is to provide 'zero-defect' logistics services with maximum efficiency through the CI of every aspect of the work processes in the firm that ultimately contributes to customer satisfaction. This first step helps top management recognize that they must be committed to and be the champion of the initiative, by personally participating in the implementation of JIT logistics to achieve the declared vision and attain the goals.

STEP TWO – IMPLEMENTATION TASKFORCE

After defining the vision, top management needs to define a set of measurable JIT logistics and performance goals and formulate the corresponding strategies to guide the firm towards achieving those goals. To this end, firms should conduct customer surveys on a regular basis to understand customer needs and identify logistics and performance objectives. The survey should cover different aspects of the workflow in the logistics activities of the firm, including storage, receiving cargo procedures, delivery cargo procedures, export procedures, import procedures, local distribution, inventory management, documentation and the overall perception of its staff. For instance, firms can request their customers to rate the importance of the workflow to them, as well as the performance of the firms in meeting their requirements. After collecting the data, importance-performance analysis can be used to provide directions for the firm to set performance objectives and strive for continuous performance improvement in JIT logistics. On the other hand, an implementation taskforce can be put together to carry out the strategies laid down by top management. The taskforce should consist of representatives who could speak for their departments and steer their departmental activities towards actions for implementing JIT logistics. The taskforce is responsible for the detailed design of the implementation of JIT logistics, defining the taskforce's scope of work, developing operations standards and assessment methods, and estimating and gathering the resources required for a successful implementation of JIT logistics.

STEP THREE – MEASUREMENT STANDARDS

This step is to develop a JIT logistics improvement model with appropriate measures to gauge outcomes and performance. The improvement model should identify the most significant outputs being produced in a firm that would determine customer satisfaction and figure out the critical performance characteristics of each of these. These JIT logistics improvement requirements should be incorporated into the business policy of the firm, which requires its staff to ensure:

- reliable inventory records;
- flawless inventory;
- services that satisfy both customers' requirements and related regulatory and legal requirements;
- continuous improvements for its logistics activities.

On the other hand, firms should set their JIT logistics objectives. The objectives may include:

- number of customer complaints decreases by 10 per cent, compared with the previous year's figure, or less than three cases;

- number of customer complaints regarding incorrect inventory records decreases by 10 per cent, compared with the previous year's figure, or less than three cases;
- 100 per cent stock take accuracy as per agreed service standard (number of discrepancy quantity/handled quantity);
- zero number in customer loss due to substandard service quality;
- increase in overall customer survey rating by 5 per cent against last survey, or above five on a seven-point scale.

The purpose of performance measurement for JIT logistics is to reveal problems so that evaluative and corrective actions can be taken, as follows: a performance review of JIT logistics should be conducted in functional departments, for example, sales and marketing, operations, accounting and technical, to reveal where performance improvement is possible, where corrections are necessary and to record actual improvements for assessment in subsequent stages.

After the performance review, the taskforce should establish written standards and procedures to govern the various aspects of the work processes in each functional department affecting customer satisfaction. The standards and procedures should provide clear guidelines and instructions to staff members on the requirements of each work process and their roles and responsibilities in meeting the requirements and the following performance pledges of the firms. Examples include:

- all the staff are to serve with sincerity and quality, without defective performance;
- all the staff are of the highest level of professionalism and follow any safety requirements;
- science management is integrated into all the commercial and logistics management activities;
- provide customers with quality service by making use of appropriate advanced information technology;
- business is operated in the most ethical way.

STEP FOUR – AWARENESS CREATION

The successful implementation of JIT logistics requires cultural change and fundamental adjustment of the behaviour of people in the firm. The organizational culture encompasses the common behaviour of all the employees in the firm with regard to their work, the firm and their relations with customers, suppliers and employees. The organizational culture can also be described as a consolidation of opinions, ideas, values, rules, behavioural patterns and norms of the people within the firm.

This step is to create quality awareness and communicate top management's vision about JIT logistics to all the employees. The purpose is to raise the personal concern and commitment of staff members to achieving the goals of JIT logistics in the firm. Managerial and supervisory staff should be imbued with the basic knowledge of JIT logistics for transmission to their subordinates. It would involve a clear explanation of the objectives of JIT logistics and educating employees in the concepts of JIT logistics, thus eliminating their fear of the implementation of JIT logistics and motivating them to participate and contribute.

STEP FIVE – MANAGER AND SUPERVISOR TRAINING

The core competences that participants in JIT logistics must develop are the ability and desire to challenge the existing structure and ways of doing business in order to generate improvements in the total supply chain through openness, shared understanding of the process and mutual respect for the benefit of all the parties involved.

Training seminars on JIT logistics should be conducted for managerial and supervisory staff. The purpose is to provide the necessary training for managers and supervisors to carry out their functions as required by JIT logistics. It is essential that all the managers and supervisors have a thorough understanding of the concepts and objectives of the implementation of JIT logistics so that they can explain them to their subordinate staff. The experiences gained from successful and failed implementations of JIT logistics in other firms should be shared.

It is also important to provide employees with the necessary education and training, equipping them with the knowledge and skills required to work effectively in the new environment. Management demonstrates its commitment through dedicating resources to improving employees. These proactive approaches share common attributes. First, they attempt to strengthen the working relationship between management and employees. Second, each demonstrates a degree of commitment to employees, either through dedicating resources as in education and training or introducing employees into what is generally regarded as areas of management concern. Third, they serve to promote equality in treatment between management and employees through the creation of organization-wide standards and sharing of information.

STEP SIX – GOAL SETTING

This step is to turn commitment into action by encouraging individual departments to set improvement plans and goals by themselves towards the organizational goal of achieving waste reduction in the logistics activities. Ultimately, the implementation efforts of JIT logistics should result in increased customer satisfaction. All the functional departments should incorporate customer satisfaction as a key objective in their work and establish goals that are specific and capable of being measured. The set goals are to be realized through concerted efforts to understand the requirements of customers, followed by effective use of company resources to meet those requirements. Examples of these initiatives are customizing services to meet specific customer needs, enhancing value for customers by offering convenient storage and distribution options, reducing order processing and delivery times, and being responsive in handling customer complaints. This step will help staff learn to think in terms of meeting goals, and to work in teams to accomplish specific tasks for improving performance in the implementation of JIT logistics.

STEP SEVEN – ERROR CAUSE REMOVAL

This step is to motivate individual staff to improve their work processes by giving them a way to communicate to management the difficulties they encountered in actually implementing JIT logistics. Listening to feedback from employees and initiating positive changes are important because the implementation of JIT logistics might cause an increase in the daily workload of the staff. The increased workload might cause some employees to become disgruntled while the lack of staff support is likely to undermine efforts to implement JIT logistics.

In this step, each individual staff member is invited to describe in a simple, one-page form any problems that hindered them in carrying out waste reduction in their work. Examples of staff inputs are: 1) the sales and marketing department make too many errors in the rate sheet or quotations; and 2) the requests from customers for the delivery of cargoes are not clear enough. This step helps the staff realize that channels exist to make their grievances known and to help them deal with the problems that arise from their efforts to pursue waste reduction in their work.

STEP EIGHT – CORRECTIVE ACTIONS

JIT logistics services require periodic preventive maintenance of work processes. Simply setting goals and identifying root causes for work problems will not automatically lead to CI of the services. Accordingly, there is a need to frequently assess the required level of performance from the work processes and to improve them as necessary to remain competitive. This step is to provide a systematic method for resolving once and for all the problems that are identified in the previous steps. To this end, the implementation taskforce should organize meetings or form work teams to hunt for specific problems in a proactive manner (for example, late delivery of cargoes by more than 1 hour), and to formulate solutions for the problems uncovered (for example, advanced planning of delivery schedules).

STEP NINE – RECOGNITION AND REWARD

Employee empowerment and staff satisfaction based on motivation, both intrinsic and extrinsic, are crucial to achieving the goals of JIT logistics. It is therefore important to reward staff participation and celebrate achievements in JIT logistics. This can help to make the implementation journey of JIT logistics more relaxed and enjoyable. In this stage, award programmes should be established to recognize staff who meet their goals or perform outstanding acts, in order to motivate their continuous commitment and support for implementing JIT logistics. As a result, firms should foster a culture of JIT in its staff by cultivating mutually supportive relations between the firm and staff, motivating and empowering staff to make decisions that prevent problems and encouraging teamwork through such means as the formation of quality improvement circles and rewarding performance equitably.

STEP TEN – CONTINUOUS IMPROVEMENT

The last step is to repeat the continuous cycle of JIT logistics. The emphasis is on making waste reduction enduring as a never-ending action in the firm. This is important because a typical implementation cycle for JIT logistics spans 12 to 18 months. During the cycle, staff turnover and market changes might adversely affect the firm's efforts to improve performance. To sustain the momentum for CI, the implementation taskforce should meet periodically to review the design of the implementation of JIT logistics and make adjustments to suit evolving market conditions. This step is important in preparing for a new implementation cycle in order to carry on with the implementation of JIT logistics.

The renewal effort helps propel the movement for JIT logistics to spiral upwards in a never-ending fashion, for the dual goals of cost reduction and service improvement in JIT logistics know no upper limits.

Summary

JIT logistics is a management approach for firms to reduce wastes and improve services in their logistics activities. The implementation requires employee involvement and cooperation from members in their supply chains. It is also essential that firms are committed with management support and resources to achieve the objectives of JIT logistics. The implementation is a long lasting and continuous journey in which the organizational and management issues should not be overlooked. In this chapter we examined the different organizational factors that are important for the successful implementation of JIT logistics. The implementation strategy, implementation plan and the performance measurement issues for JIT logistics were fully discussed. Finally, a ten-step approach for the implementation of JIT logistics was suggested to conclude this book. While this book only gives an introduction to JIT logistics, firms will find the concepts and principles of JIT logistics useful to embark on their journey for JIT logistics and attain the 7Rs in managing their logistics activities.

Bibliography

Ahire, S.L., Golhar, D.Y. and Waller, M.A. (1996) Development and validation of TQM implementation constructs, *Decision Sciences*, Vol. 27, No. 1, pp. 23–56

Aichlmayr M. (2001) The future of JIT – time will tell, *Transportation & Distribution*, Vol. 42, No. 12, pp. 18–23.

Alles, M., Amershi A., Data, S. and Sarkar, R. (2000) Information and incentive effects of inventory in JIT production, *Management Science*, Vol. 46, No. 12, pp. 1528–1544.

Alles, M., Datar, S.M. And Lambert, R.A. (1995) Moral hazard and management control in just-in-time settings, *Journal of Accounting Research*, Vol. 33 (suppl.), pp. 177–204.

Alternburg, K., Griscom, D., Hart, J., Smith, F. and Wohler (1999) Just-in-time logistics support for the automobile industry, *Production and Inventory Management Journal*, Vol. 40, No. 2, pp. 59–66.

Ansari, A. and Modarress, B. (1990) *Just-in-Time Purchasing*, The Free Press, New York, NY.

Asubonteng, P., McCleary, K. and Swan, J. (1996) SERVQUAL revisited: a critical review of service quality, *The Journal of Services Marketing*, Vol. 10, No. 6, pp. 62–81.

Bagchi, P. (1997) Logistics benchmarking as a competitive strategy: some insights, *Logistics Information Management*, Vol. 10, No. 1, pp. 28.

Bagchi, P., Raghunathan, T. and Bardi (1987) The implications of just-in-time inventory policies on carrier selection, *Logistics and Transportation Review*, Vol. 23, No. 4, pp. 373–384.

Baker, W.E. and Sinkula, J.M. (1999) The synergistic effect of market orientation and learning orientation on organizational performance, *Journal of Academy of Marketing Science*, Vol. 27, No. 4, pp. 411–427.

Balakrishnan, R., Linsmeier, T.J. and Venkatachalam, M. (1996) Financial benefits from JIT adoption: effects of customer concentration and cost structure, *The Accounting Review*, Vol. 71, No. 2, pp. 183–205.

Ballou, R.H. (1987) *Basic Business Logistics: Transportation, Materials Management, and Physical Distribution*, 2nd edn, Prentice-Hall International Inc., Englewood Cliffs, NJ.

Ballou, R.H. (2004) *Business Logistics/Supply Chain Management: Planning, Organizing, and Controlling the Supply Chain*, 5th edn, Pearson Education, Prentice Hall, Upper Saddle River, New York.

Barlow, G.L. (2002) Just-in-time: Implementation within the hotel industry – a case study, *International Journal of Production Economics*, Vol. 80, No. 2, pp. 155–167.

Barnerjee, A. and Kim, S. (1995) An integrated JIT inventory model, *International Journal of Operations & Production Management*, Vol. 15, No. 9, pp. 237–244.

Beer, M. (2003) Why total quality management programs do not persist: the role of management quality and implications for leading a TQM transformation, *Decision Sciences*, Vol. 34, No. 4, pp. 623–643.

Benito, J., Gonzalez I. and Spring, M. (2000) Complementarities between JIT purchasing practices: An economic analysis based on transaction costs, *International Journal of Production Economics*, Vol. 67, No. 3, pp. 279–293.

Benson, R. J. (1986) JIT: Not just for the factory, Proceedings from the 29th Annual International Conference for the American Production and Inventory Control Society, St Louis, MO, No. 20–24, October, pp. 370–374.

Berman, S.L., Wicks, A.C., Kotha, S. and Jones, T.M. (1999) Does stakeholder orientation matter? The relationship between stakeholder management models and firm financial performance, *Academy of Management Journal*, Vol. 42, No. 5, pp. 488–507.

Bichou, K., Lai, K.H., Lun, V.Y.H. and Cheng, T.C.E. (2007). A quality management framework for liner shipping companies to implement the 24-hour advance vessel manifest rule, *Transportation Journal*, Vol. 46, No. 1, pp. 5–21.

Billesbach, T. and Hayen, R. (1994) Long-term impact of just-in-time on inventory performance measures, *Production and Inventory Management Journal*, Vol. 35, No. 1, pp. 62–67.

Bookbinder, J. H. and Dilts, D. M. (1989) Logistics information systems in a just-in-time environment, *Journal of Business Logistics*, Vol. 10, No. 1, pp. 50–67.

Bowen, F.E., Cousins, P.D., Lamming, R.C. and Faruk, A.C. (2001) The role of supply management capabilities in green supply, *Production and Operations Management*, Vol. 10, No. 2, pp. 174–89.

Bowersox, D., Closs, D. and Cooper, M. (2002) *Supply Chain Logistics Management*, McGraw-Hill, New York.

Bowersox, D., Closs, D. and Helferich, O. (1986) *Logistical Management: A Systems Integration of Physical Distribution, Manufacturing Support, and Materials Procurement*, 3rd edn, Macmillan Publishing Company, Boston, Massachusetts.

Boyle, M. (2003) Wal-Mart keeps the change, *Fortune*, Vol. 148, No. 19, p. 46.

Brinker, B. (2000) *Guide to Cost Management*, John Wiley & Sons, Inc, New York.

Callen, J.L., Fader, C. and Krinsky, I. (2000) Just in time: A cross sectional plant analysis, *International Journal of Production Economics*, Vol. 63, No. 3, pp. 277–301.

Callen, J.L., Morel, M. and Fader, C. (2005) Productivity measurement and the relationship between plant performance and JIT intensity, *Contemporary Accounting Research*, Vol. 22, No. 2, pp. 271–309.

Canel, C., Rosen, D. and Anderson, E. (2000) Just-in-time is not just for manufacturing: A service perspective, *Industrial Management & Data Systems*, Vol. 100, No. 2, pp. 51–60.

Cannon, J.P. and Homburg, C. (2001) Buyer-supplier relationships and customer firm costs, *Journal of Marketing*, Vol. 65, No.1, pp. 29–43.

Chandy, R.K., and Tellis, G.J. (1998) Organizing for radical product innovation: The overlooked role of willingness to cannibalize, *Journal of Marketing Research*, Vol. 35, No. 4, pp. 474–487.

Chapman, S.N. and Carter, P.L. (1990) Supplier/customer inventory relationships under just-in-time, *Decision Sciences*, Vol. 21, No. 1, pp. 35–51.

Chappell, G., Ginsburg, L., Schmidt, P., Smith, J. and Tobolski, J. (2002) Auto-ID on Demand: The Value of Auto-ID Technology in Consumer Packaged Goods Demand Planning. *Auto-ID Center*, pp. 1–25

Chase, R.B., Jacobs, F.R. and Aquilano, N.J. (2006) *Operations Management for Competitive Advantage*, 11th edn, McGraw-Hill.

Cheng, T.C.E. and Podolsky, S. (1996) *Just-in-time Manufacturing: An Introduction*, 2nd edn, Chapman & Hall, London.

Cheng, T.C.E. and Wu, Y.N. (2005) The impact of information sharing in a two-level supply chain with multiple suppliers, *Journal of the Operational Research Society*, Vol. 56, No. 10, pp. 1159–1165.

Chopra, S. and Meindl, P. (2007) *Supply Chain Management: Strategy, Planning, and Operation*, 3rd edn, Prentice Hall, Upper Saddle River, New York.

Christopher, M. (2005) *Logistics and Supply Chain Management: Creating Value-Adding Networks*, 3rd edn, Financial Times Prentice Hall, New York.

Christopher, M. and Lee, H.L. (2004) Mitigating supply chain risk through improved confidence, *International Journal of Physical and Distribution Management*, Vol. 34, No. 5, pp. 388–396.

Christopher, M. and Towill, D.R. (2002) Developing market specific supply chain strategies, *The International Journal of Logistics Management*, Vol. 13, No. 1, pp. 1–14.

Clouse, V. and Gupta, Y. (1990) Just-in-time and the trucking industry: Implications of the motor carrier act, *Production and Inventory Management Journal*, Fourth Quarter, pp. 7–12.

Cooper, M.C., Lambert, D.M. and Pagh, J.D. (1997) Supply chain management: More than a new name for logistics, *International Journal of Logistics Management*, Vol. 8, No.1, pp. 1–14.

Cohen, W. M. and Levinthal, D. A. (1990) Absorptive capacity: A new perspective on learning and innovation, *Administrative Science Quarterly*, Vol.35, No. 1, pp. 867–888.

Coyle, J., Bardi, E. And Langley, Jr. C. (2003) *The Management of Business Logistics: A Supply Chain Perspective*, 7th edn, South-western Thomson Learning, Cincinnati, Ohio.

Croxton, K.L., Carcia-Dastugue, S.J. and Lambert, D.M. (2001) The supply chain management processes, *International Journal of Logistics Management*, Vol. 12, No. 2, pp. 13–36.

Cusumano, M.A. and Takeishi, A. (1991) Supplier relations and management: A survey of Japanese, Japanese-transplant, and U.S. auto plants, *Strategic Management Journal*, Vol. 12, No. 8, pp. 563–588.

Davy, J.A., White, R.E., Merritt, N.J. and Gritzmacher, K. (1992) A derivation of the underlying constructs of just-in-time management systems, *Academy of Management Journal*, Vol. 35, No. 3, pp. 653–671.

De Toni, A. and Nassimbeni, G. (2000) Just-in-time purchasing: An empirical study of operational practices, supplier development and performance, *Omega*, Vol. 28, No. 6, pp. 631–651.

de Treville, S. and Antonakis, J. (2006) Could lean production job design be intrinsically motivating? Contextual, configurational, and levels-of-analysis issues, *Journal of Operations Management*, Vol. 24, No. 2, pp. 99–123.

DiMaggio, P.J. and Powell, W.W. (1983) The iron cage revisited: Institutional isomorphism and collective rationality in organizational fields, *American Sociological Review*, Vol. 48, No. 2, pp. 147–160.

Dong, Y., Carter, C.R. and Dresner, M.E. (2001) JIT purchasing and performance: An exploratory analysis of buyer and supplier perspective, *Journal of Operations Management*, Vol. 19, No. 4, pp. 471–483.

Dowlatshahi, S. (2005) A strategic framework for the design and implementation of remanufacturing operations in reverse logistics, *International Journal of Production Research*, Vol. 43, No. 16, pp. 3455–3480.

Dreyfus, L.P., Ahire, S.L. and Ebrahimpour, M. (2004) The impact of just-in-time implementation and ISO 9000 certification on total quality management, *IEEE Transactions on Engineering Management*, Vol. 51, No. 2, pp. 125–141.

Droge, C. and Germain, R. (1998) The just-in-time inventory effect: Does it hold under different contextual, environmental, and organizational conditions? *Journal of Business Logistics*, Vol. 19, No. 2, pp. 53–71.

Drucker, P.F. (1987) Worker hands bound by tradition, *Wall Street Journal*, August, 2, 18.

Dwyer, F.R., Schurr, P.H. and Oh, S. (1987) Developing buyer-seller relationships. *Journal of Marketing*, Vol. 51, No. 2, pp. 11–28.

Epps, R. (1995) Just-in-time inventory management: Implementation of a successful program, *Review of Business*, Vol. 17, No. 1, pp. 40–44.

Feitzinger, E. and Lee, H.L. (1997) Mass customization at Hewlett-Packard: The power of postponement, *Harvard Business Review*, Vol. 75, No. 1, pp. 116–121.

Fielder, K., Galletly, J.E. and Bicheno, J. (1993) Expert advice for JIT implementation, *International Journal of Operations and Production Management*, Vol. 13, No. 6, pp. 23–30.

Finkenzeller, K. (2003) *RFID Handbook: Radio-Frequency Identification Fundamentals and Applications*, John Wiley & Son, Ltd., England.

Fitzsimmons, J.A. (2006) *Service Management: Operations, Strategy, and Information Technology*, 5th edn, McGraw-Hill, New York.

Flapper, S.D.P., van Nunen, Jo A.E.E. and Wassenhove, L.N.V. (2005) *Managing Closed-loop Supply Chains*, Springer, New York.

Fliedner, G. and Vokurka, R.J. (1997) Agility: Competitive weapon of the 1990s and beyond? *Production and Inventory Management Journal*, Vol. 38, No. 3, pp. 19–24.

Frazelle, E. (2002) *Supply Chain Strategy: The Logistics of Supply Chain Management*, McGraw-Hill., New York.

Frazier, G.L., Spekman, R.E. and O'Neal, C.R. (1988) Just-in-time exchange relationships in industrial markets, *Journal of Marketing*, Vol. 52, No.4, pp. 52–67.

Fullerton, R.R. and McWatters, C.S. (2001) The production performance benefits from JIT implementation, *Journal of Operations Management*, Vol. 19, No. 1, pp. 81–96.

Fullerton, R.R. and McWatters, C.S. (2002) The role of performance measures and incentive systems in relation to the degree of JIT implementation, *Accounting, Organizations and Society*, Vol. 27, No. 8, pp. 711–735.

Fynes, B. and Voss, C. (2002) The moderating effect of buyer-supplier relationships on quality practices and performance, *International Journal of Operations & Production Management*, Vol. 22, No. 6, pp. 589–613.

Garvin, D.A. (1983) Quality on the line, *Harvard Business Review*, Vol. 61, No. 5, pp. 64–75.

Garvin, D.A. (1987) Competing on the eight dimensions of quality, *Harvard Business Review*, Vol. 65, No. 6, pp. 101–109.

Gélinas, R. (1999) The just-in-time implementation project, *International Journal of Project Management*, Vol. 17, No. 3, pp. 171–179.

Gentry, J. (1996) The role of carriers in buyer-supplier strategic partnerships: A supply chain management approach, *Journal of Business Logistics*, Vol. 17, No. 2, pp. 33–55.

Germain, R. and Droge, C. (1997) Effect of just-in-time purchasing relationships on organizational design, purchasing department configuration, and firm performance, *Industrial Marketing Management*, Vol. 26, No. 2, pp. 115–125.

Germain, R., Droge, C. and Daugherty, P. (1994) The effect of just-in-time selling on organizational structure: an empirical investigation, *Journal of Marketing Research*, Vol. 31, No. 4, pp. 471–483.

Germain, R. Droge, C. and Spears, N. (1996) The implications of just in time for logistics organization management and performance, *Journal of Business Logistics*, Vol. 17, No. 2, pp. 19–34.

Giunipero, L. and O'Neal, C. (1988) Obstacles to JIT procurement, *Industrial Marketing Management*, Vol. 17, No.1, pp. 35–41.

Goldsby, T.J. and Garcia-Dastugue, S.J. (2003) The manufacturing flow management process, *International Journal of Logistics Management*, Vol. 14, No. 2, pp. 33–52.

Gonzalez-Benito, J. (2002) Effects of the characteristics of the purchased products in JIT purchasing implementation, *International Journal of Operations and Production Management*, Vol. 22, No. 8, pp. 868–886.

Gourdin, K. (2006) *Global Logistics Management: A Competitive Advantage for the 21st Century*, Blackwell Publishers, Malden, MA.

Graham, I. (1988) Japanisation as mythology, *Industrial Relations Journal*, Vol. 29, No. 1, pp. 69–75.

Grapentine, T. (1998) The history and future of service quality assessment – connecting customer needs and expectations to business processes, *Marketing Research*, Vol. 10, No. 4, pp. 5–20.

Green, K.W. and Inman, R.A. (2005) Using a just-in-time selling strategy to strengthen supply chain linkages, *International Journal of Production Research*, Vol. 43, No. 15, pp. 3437–3453.

Green, K.W., Inman, R.A. and Brown, G. (2008) Just-in-time selling construct: Definition and measurement, *Industrial Marketing Management*, Vol. 37, No. 2, pp. 131–142.

Gunasekaran, A. (1999) Just-in-time purchasing: An investigation for research and applications, *International Journal of Production Economics*, Vol. 59, No. 1–3, pp. 77–84.

Gunasekaran, A., Lai, K.H. and Cheng, T.C.E. (2008) Responsive supply chain: A competitive strategy in the networked economy, *Omega*, Vol. 36, No. 4, pp. 549–564.

Hall, E. Jr. (1989) Just-in-time management: A critical assessment, *The Academy of Management Executive*, Vol. 3, No. 4, pp. 315–318.

Hall, R. W. (1987) *Attaining Manufacturing Excellence: Just in Time, Total Quality, Total People Involvement*, Dow Jones-Irwin, Illinois.

Hallihan, A., Sackett, P. and Williams, G.M. (1997) JIT manufacturing: The evolution to and implementation model founded in current practice, *International Journal of Production Research*, Vol. 35, No. 4, pp. 901–920.

Ham C., Johnson W., Weinstein A., Plank R. and Johnson P. (2003) Gaining competitive advantages: analyzing the gap between expectations and perceptions of service quality, *International Journal of Value-Based Management*, Vol. 16, No. 2, pp. 197–203.

Hamel, G. and Prahalad, C.K. (1989) Strategic intent, *Harvard Business Review*, Vol. 67, No. 3, pp. 63–76.

Hay, E.J. (1988) *The Just-in-Time Breakthrough*, John Wiley and Sons, New York.

Heiko, L. (1989) Some relationship between Japanese culture and just-in-time, *Academy of Management Executive*, Vol. 3, No. 4, pp. 319–321.

Heskett, J., Glaskowsky, N.A. and Ivie, R.M. (1973) *Business Logistics*, New York, Ronald Press Company.

Hill, T. (2000) The Wal-Mart case study. In: *Operations Management Strategic Concepts and Management Analysis*, Macmillan Press, Basingstoke, UK, p. 49.

Hopp, W.J. and Spearman, M.L. (2004) To pull or not to pull: What is the question? *Manufacturing & Service Operations Management*, Vol. 6, No.2, pp. 133–148.

Houlihan, J.B. (1985) International supply chain management, *International Journal of Physical Distribution and Materials Management*, Vol. 15, No. 1, pp. 22–38.

Hult, G.T.M. and Ketchen, D.J. (2001) Does market orientation matter? A test of the relationship between positional advantage and performance, *Strategic Management Journal*, Vol. 22, No. 9, pp. 899–906.

Hurley, S. and Whybark (1999) Comparing JIT approaches in a manufacturing cell, *Production and Inventory Management Journal*, Vol. 40, No. 2, pp. 32–37.

Huson, M. and Nanda, D. (1995) The impact of just in time manufacturing on firm performance in the U.S., *Journal of Operations Management*, Vol. 12, Vol. 3, pp. 297–310.

Hwang, Y. D., Lin Y. C. and Lyu, J. (2008) The performance evaluation of SCOR sourcing process – the case study of Taiwan's TFT-LCD industry, *International Journal of Production Economics*, Vol. 115, No. 2, pp. 411–423.

Im, J.H. and Lee, S.M. (1989) Implementation of just-in-time systems in U.S. manufacturing firms, *International Journal of Operations and Production Management*, Vol. 9, No. 1, pp. 5–14.

Inman, R.A. and Mehra, S. (1993) Financial justification of JIT implementation, *International Journal of Operations and Production Management*, Vol. 13, No. 4, pp. 32–39.

Jayaraman, V., Guide, V.D.R. and Srivastava, R. (1999) Closed-loop logistics model for remanufacturing, *Journal of the Operational Research Society*, Vol. 50, No. 5, pp. 497–508.

Jayaram, J. and Vickery, S.K. (1998) Supply-based strategies, human resource initiatives, procurement leadtime, and firm performance, *International Journal of Purchasing and Materials Management*, Vol. 34, No. 1, pp. 12–23.

Johnson, J., Wood, D., Wardlow, D. and Murphy, P. (1999) *Contemporary Logistics*, 7th edn, Prentice-Hall, Upper Saddle River, New York.

Kaneko, J. and Nojiri, W. (2008) The logistics of Just-in-Time between parts suppliers and car assemblers in Japan, *Journal of Transport Geography*, Vol. 16, No. 3, pp. 155–173.

Kannan, V.R. and Tan, K.C. (2005) Just in time, total quality management, and supply chain management: Understanding their linkages and impact on business performance, *Omega*, Vol. 33, No. 2, pp. 153–162.

Kaplan, R.S. and Norton, D.P. (1992) The balanced scorecard – measures that drive performance, *Harvard Business Review*, Vol. 70, No. 1, pp. 71–79.

Karlson, C. and Awstorm, P. (1996) Assessing changes towards lean production, *International Journal of Operations and Production Management*, Vol. 16, No. 2, pp. 46–64.

Kaynak, H. and Hartley, J.L. (2006) Using replication research for just-in-time purchasing, *Journal of Operations Management*, Vol. 24, No. 6, pp. 868–892.

King, J.L., Gurbaxani, V., Kraemer, K.L., McFarlan, F.W., Raman, K.S. and Yap, C.S. (1994) Institutional factors in information technology innovation, *Information Systems Research*, Vol. 5, No. 2, pp. 139–169.

Kinney, M.R. and Wempe, W.F. (2002) Further evidence on the extent and origins of JIT's profitability effects, *The Accounting Review*, Vol. 77, No. 1, pp. 203–225.

Kohli, A.K. and Jaworski, B.J. (1990) Market orientation: The construct, research propositions and consequences, *Journal of Marketing*, Vol. 54, No. 2, pp. 1–18.

Krajewski, L.J. and Ritzman, L.P. (2005) *Operations Management: Processes and Value Chains*, 7th edn, Pearson/Prentice-Hall, New York.

Krause, D.R., Pagell, M. and Curkovic, S. (2001) Toward a measure of competitive priorities for purchasing, *Journal of Operations Management*, Vol. 19, No. 2, pp. 497–512.

Krikke, H., Blanc, L.L. and van de Velde, S. (2004) Product modularity and the design of closed-loop supply, *California Management Review*, Vol. 48, No. 2, pp. 6–19.

Koufteros, X.A., Cheng, T.C.E. and Lai, K.H. (2007a) Black-box and gray-box supplier integration in product development: Antecedents, consequences and the moderating role of firm size, *Journal of Operations Management*, Vol. 25, No. 4, pp. 847–870.

Koufteros, X.A., Nahm, A.Y., Cheng, T.C.E. and Lai, K.H. (2007b) An empirical assessment of a nomological network of constructs: From customer orientation to pull production and performance, *International Journal of Production Economics*, Vol. 106, No. 2, pp. 468–492.

Lai, K.H. (2003) Market orientation in quality-oriented organizations and its impact on their performance, *International Journal of Production Economics*, Vol. 84, No. 1, pp. 17–34.

Lai, K.H. (2004) Service capability and performance of logistics service providers, *Transportation Research Part E*, Vol. 40, No. 5, pp. 385–399.

Lai, K.H., Bao, Y. and Li, X. (2008a) Channel relationship and business uncertainty: Evidence from the Hong Kong market, *Industrial Marketing Management*, Vol. 37, No. 6, pp. 713–724.

Lai, K.H. and Cheng, T.C.E. (2003a) Initiatives and outcomes of quality management implementation across industries, *Omega*, Vol. 31, No. 2, pp. 141–154.

Lai, K.H. and Cheng, T.C.E. (2003b) Supply chain performance in transport logistics: An assessment by service providers, *International Journal Logistics: Research and Applications*, Vol. 6, No. 3, pp. 151–164.

Lai, K. H. and Cheng, T. C. E. (2004) A study of the freight forwarding industry in Hong Kong, *International Journal of Logistics: Research and Applications*, Vol. 7, No. 2, pp. 71–84.

Lai, K.H., Cheng, T.C.E. and Yeung, A.C.L. (2005a) Relationship stability and supplier commitment to quality, *International Journal of Production Economics*, Vol. 96, No. 3, pp. 397–410.

Lai, K.H., Lau, G. and Cheng, T.C.E. (2004) Quality management in the logistics industry: An examination and a ten-step approach for quality implementation, *Total Quality Management and Business Excellence*, Vol. 15, No. 2, pp. 147–160.

Lai, K.H., Ngai E.W.T. and Cheng, T.C.E (2002) Measures for evaluating supply chain performance in transport logistics, *Transportation Research Part E*, Vol. 38, No. 6, pp. 439–456.

Lai, K.H., Ngai, E.W.T. and Cheng, T.C.E. (2005b) Information technology adoption in Hong Kong's logistics industry, *Transportation Journal*, Vol. 44, No. 4., pp. 1–9.

Lai, K.H., Wong, C.W.Y. and Cheng, T.C.E. (2006a) Institutional isomorphism and the adoption of information technology for supply chain management, *Computers in Industry*, Vol. 57, No. 3, pp. 93–98.

Lai, K.H., Wong, C.W.Y. and Cheng, T.C.E. (2008b) A coordination-theoretic investigation of the impact of electronic integration on logistics performance, *Information & Management*, Vol. 45, No.1, pp. 10–20.

Lai, K.H., Wong, C.W.Y., Cheng, T.C.E. and Yeung, A.C.L. (2006b) Antecedents and consequences of electronic product code adoption and its implications for supply chain management: A framework and propositions for future research, *Maritime Economics and Logistics*, Vol. 8, No. 4, pp. 311–330.

Lambert, D.M. (2001) The supply chain management and logistics controversy, in Brewer, A.M., Button, K.J., Hensher, D.A. (Eds), *Handbook of Logistics and Supply Chain Management*, Pergamon, Oxford, pp. 99–126.

Lambert, D.M., Cooper, M.C. and Pagh, J.D. (1998) Supply chain management: Implementation issues and research opportunities, *International Journal of Logistics Management*, Vol. 9, No. 2, pp. 1–19.

Lamming, R. (1993) *Beyond Partnership – Strategies for Innovation and Lean Supply*, Prentice-Hall, New York.

Lamming, R. and Hampson, J. (1996) The environment as a supply chain management issue, *British Journal of Management*, Vol. 7, No. 1, pp. 45–62.

Lee, H.L. (2000) Creating value through supply chain integration, *Supply Chain Management Review*, Vol. 4, No. 4, pp. 30–36.

Lee, H.L. (2004) Simple theories for complex logistics, *Optimize*, July, pp. 42–46.

Lee, H.L. and Billington, C. (1992) Managing supply chain inventory: Pitfalls and opportunities, *Sloan Management Review*, Vol. 33, No. 3, pp. 65–73.

Lee, H.L., Padmanabhan, V. and Whang, S. (1997) The bullwhip effect in supply chains, *Sloan Management Review*, Vol. 38, No. 3, pp. 93–102.

Lee, H.L. and Whang, S. (2005) Higher supply chain security with lower cost: Lessons from total quality management, *International Journal of Production Economics*, Vol. 96, No. 3, pp. 289–300.

Lee, S.M. and Ebrahimpour, M. (1984) Just in time production system: Some requirements for implementation, *International Journal of Operations and Production Management*, Vol. 4, No. 4, pp. 3–15.

Li, J., Cheng, T.C.E. and Wang, S.Y., (2007) Analysis of postponement strategy for perishable items by EOQ-based models, *International Journal of Production Economics*, Vol. 107, No. 1, pp. 31–38.

Li, J., Wang, S.Y. and Cheng, T.C.E. (2008) Analysis of postponement strategy by EPQ-based models with planned backorders, *Omega*, Vol. 36, No. 5, pp. 777–788.

Lines, R. (2004) Influence of participation in strategic change: Resistance, organizational commitment and change goal achievement, *Journal of Change Management*, Vol. 4, No. 3, pp. 193–216.

Lu, C.S., Lai, K.H. and Cheng, T.C.E. (2005) An evaluation of web site services in liner shipping in Taiwan, *Transportation*, Vol. 32, No. 3, pp. 293–318.

Lu, C.S., Lai, K.H. and Cheng, T.C.E. (2006) Adoption of Internet services in liner shipping – an empirical study of shippers in Taiwan, *Transport Reviews*, Vol. 26, No. 2, pp. 189–206.

Lun, Y.H.V., Lai, K.H. and Cheng, T.C.E. (2009) *Container Transport Management*, Shipping and Transport Logistics Book Series, Inderscience, Geneva, Switzerland.

Lun, Y.H.V., Wong, C.W.Y., Lai, K.H. and Cheng, T.C.E. (2008) Institutional perspective on the adoption of technology for container transport security enhancement, *Transport Reviews*, Vol. 28, No. 1, pp. 21–33.

Maloni, M. and Benton, W.C. (2000) Power influences in the supply chain, *Journal of Business Logistics*, Vol. 21, No. 1, pp. 49–73.

Martilla, J.A. and James, J.C. (1977) Importance-performance analysis, *Journal of Marketing*, Vol. 41, No. 1, pp. 77–79.

Massey, A.P., Montoya-Weiss, M.M. and Holcom, K. (2001) Re-engineering the customer relationship: Leveraging knowledge assets at IBM, *Decision Support Systems*, Vol. 32, No. 2, pp. 155–170.

Matsui, Y. (2007) An empirical analysis of just-in-time production in Japanese manufacturing companies, *International Journal of Production Economics*, Vol. 108, No. 1–2, pp. 153–164.

McDougall G. and Levesque T. (2000) Customer satisfaction with services: Putting perceived value into the equation, *Journal of Services Marketing*, Vol. 14, No. 5, pp. 392–410.

McEvily, S.K. and Chakravarthy, B. (2002) The persistence of knowledge-based advantage: An empirical test for product performance and technological knowledge, *Strategic Management Journal*, Vol. 23, No. 4, pp. 285–305.

McGillivray, G. (2000) Commercial risk under JIT (Just-in-time efficiency supply process), *Canadian Underwriter*, Vol. 67, No. 1, pp. 26.

Mehra S. and Inman, A. (1992) Determining the critical elements of just-in-time implementation, *Decision Sciences*, Vol. 23, No. 1, pp. 160–174.

Mentzer, J.T., Flint, D.J. and Hult, G.T.M. (2001) Logistics service quality as a segment-customized process, *Journal of Marketing*, Vol. 65, No. 4, pp. 82–104.

Mia, L. (2000) Just-in-time manufacturing, *Accounting and Business Research*, Vol. 30, No. 2, pp. 137–151.

Mollenkopf, D.A. and Closs, D.J. (2005), The hidden value in reverse logistics, *Supply Chain Management Review*, Vol. 9, No. 5, pp. 34–43.

Nagel, R. N. and Bhargava, P. (1994) Agility: The ultimate requirement for world-class manufacturing performance, *National Productivity Review*, Vol. 13, No. 3, pp. 331–340.

Narasimhan, R. and Carter, J.R. (1998) *Environmental Supply Chain Management*, The Center for Advanced Purchasing Studies, Arizona State University, Tempe, AZ.

Narasimhan, R. and Jayaram, J. (1998) Causal linkages in supply chain management: An exploratory study of North American manufacturing firms, *Decision Sciences*, Vol. 29, No. 3, pp. 579–605.

Ngai, E.W.T., Cheng, T.C.E., Au, S. and Lai, K.H. (2007a) Mobile commerce integrated with RFID technology in a container depot, *Decision Support Systems*, Vol. 43, No. 1, pp. 62–76.

Ngai, E.W.T., Cheng, T.C.E., Lai, K.H., Chai, P.Y.F., Choi, Y.S. and Sin, R.K.Y. (2007b) Development of an RFID-based traceability system: A case study in aircraft engineering company, *Production and Operations Management*, Vol. 16, No. 5, pp. 554–568.

Ngai, E.W.T., Lai, K.H. and Cheng, T.C.E. (2008) Logistics information systems: The Hong Kong experience, *International Journal of Production Economics*, Vol. 113, No. 1, pp. 223–234.

Oakland, J.S. (1993) *Total Quality Management: The Route to Improving Performance*, 2nd edn, Nichols Publishing Company, New Jersey,.

Ohno, T. (1982) How the Toyota production system was created, *Japanese Economic Studies*, Vol. 10, No. 4, pp. 83–101.

Pagell, M., Yang, C.L, Krumwiede, D.W. and Sheu, C. (2004) Does the competitive environment influence the efficacy of investment in environmental management?, *Journal of Supply Chain Management*, Vol. 40, No. 3, pp. 30–39.

Parasuraman, A., Zeithaml, V. and Berry, L. (1985) A conceptual model of service quality and its implications for future research, *Journal of Marketing*, Vol. 49, No. 4., pp. 41–50.

Parasuraman, A., Zeithaml, V. and Berry, L. (1997) Listening to the customer – the concept of a service-quality information system, *Sloan Management Review*, Vol. 38, No. 3, pp. 65–76.

Parasuraman, A., Zeithaml, V. and Berry, L. (1988) SERVQUAL: A multiple-item scale for measuring consumer perceptions of service quality, *Journal of Retailing*, Vol. 64, No. 1, pp. 12–40.

Pine, J.B. (1993) *Mass Customization: The Frontier in Business Competition*, Harvard Business School Press, Boston, Massachusetts.

Plenert, G. (1999) Focusing material requirement planning (MRP) towards performance, *European Journal of Operational Research*, Vol. 119, No. 1, pp. 91–99.

Polito, T. and Watson, K. (2006) Just-in-time under fire: The five major constraints upon JIT practices, *The Journal of American Academy of Business*, Vol. 9, No. 1, pp. 8–13.

Porter, M. (1985) *Competitive Advantage Creating and Sustaining Superior Performance*, The Free Press, New York.

Power, D. and Sohal, A. (1997) An examination of the literature relating to issues affecting the human variable in just-in-time environments, *Technovation*, Vol. 17, Nos. 11/12, pp. 649–666.

Prince, J. and Kay, J.M. (2003) Combining lean and agile characteristics: creation of virtual groups by enhanced production flow analysis, *International Journal of Production Economics*, Vol. 85, No. 3, pp. 305–318.

Rogers, E.M. (1983) *The Diffusion of Innovations*, 3rd edn, Free Press, New York.

Russell, D.M. and Saldanha, J.P. (2003) Five tenets of security-aware logistics and supply chain operation, *Transportation Journal*, Vol. 42, No. 4, pp. 44–54.

Sakakibara, S., Flynn, B.B. and Schroeder, R.G. (1993) A framework and measurement instrument for just-in-time manufacturing, *Production and Operations Management*, Vol. 2, No. 3, pp. 177–194.

Salaheldin, S. (2005) JIT implementation in Egyptian manufacturing firms: Some empirical evidence, *International Journal of Operations and Production Management*, Vol. 25, No. 4, pp. 354–370.

Scannell, T.V. Vickery, S.K. and Droge, C.L. (2000) Upstream supply chain management and competitive performance the automotive supply industry, *Journal of Business Logistics*, Vol. 21, No. 1, pp. 23–48.

Schniederjans, M.J. (1993) *Topics in Just-in-Time Management*, Allyn and Bacon, Boston, Massachusetts.

Schonberger, R.J. (1982a) *Japanese Manufacturing Techniques: Nine Hidden Lessons in Simplicity*, New York, NY, Free Press.

Schonberger, R.J. (1982b) Some observations on advantages and implementation issues of just-in-time production systems, *Journal of Operations Management*, Vol. 3, No. 1, pp. 1–11.

Schonberger, R.J. (1986) *World Class Manufacturing: The Lessons of Simplicity Applied*, New York, NY, Free Press.

Seuring, S. (2004) Industrial ecology, life cycles, supply chains: Differences and interrelations, *Business Strategy and the Environment*, Vol. 13, No. 5, pp. 306–319.

Sheffi, Y. (2005) *The Resilient Enterprise: Overcoming Vulnerability for Competitive Advantage*, The MIT Press, Cambridge, Massachusetts.

Simchi-Levi, D., Kaminsky, P. and Simchi-Levi, E. (2008) *Designing and Managing the Supply Chain: Concepts, Strategies, and Case Studies*, 3rd edn, McGraw-Hill/ Irwin, Boston.

Singh, N., Lai, K.H. and Cheng, T.C.E. (2007) Inter-organizational perspective on IT-enabled supply chains, *Communications of the ACM*, Vol. 50, No. 1, pp. 59–65.

Sitkin, S.B., Sutcliffe, K.M. and Schroeder, R.G. (1994), Distinguishing control from learning in total quality management, *Academy of Management Review*, Vol. 19, No. 3, pp. 537–564.

Sohal, A.S., Ramsay, L. and Samson, D. (1993) JIT manufacturing: Industry analysis and a methodology for implementation, *International Journal of Operations and Production Management*, Vol. 13, No. 7, pp. 22–56.

Sowinski, L. and Orton, C. (2001) Move over, JIT, *World Trade*, Vol. 14, No. 4, pp. 54–56.

Srinivasan, R., Lilien, G.L. and Rangaswamy, A. (2002) Technological opportunism and radical technology adoption: An application to e-business, *Journal of Marketing*, Vol. 66, No. 3, pp. 47–60.

Stank, T.P., Crum, M. and Arango, M. (1999) Benefits of interfirm coordination in food industry supply chains, *Journal of Business Logistics*, Vol. 20, No. 2, pp. 21–41.

Stank, T.P., Goldsby, T.J., Vickery, S.K. and Savitskie, K. (2003) Logistics service performance: estimating its influence on market share, *Journal of Business Logistics*, Vol. 24, No. 1, pp. 27–56.

Stanley, L.L. and Wisner, J.D. (2001) Service quality along the supply chain: Implications for purchasing, *Journal of Operations Management*, Vol. 19, No. 3, pp. 287–306.

Stock, J.R. and Lambert, D.M. (2001) *Strategic Logistics Management*, 4th edn, McGraw-Hill, New York.

Teece, D.J. (1986) Profiting from technological innovation: implications for integration, collaboration, licensing, and public policy, *Research Policy*, Vol. 15, No. 6, pp. 285–305.

Teo, T. S. H., Lin, S. and Lai, K. H. (2009) Adopters and non-adopters of e-procurement in Singapore: An empirical study, Omega, Vol. 37, No. 5, pp. 972–987.

Thierry, M., Salomon, M., van Nunen, J. and van Wassenhove, L.N. (1995) Strategic issues in product recovery management, *California Management Review*, Vol. 37, No. 2, pp. 114–135.

Thompson, V.A. (1965) Bureaucracy and innovation, *Administrative Science Quarterly*, Vol. 10, No. 1, pp. 1–20.

Tibben-Lembke, R.S. (2004) Strategic use of the secondary market for retail consumer, *California Management Review*, Vol. 48, No. 2, pp. 90–104.

Toffel, M. (2004) Strategic management of product recovery, *California Management Review*, Vol. 48, No. 2, pp. 120–141.

Tornatzky, L.G. and Fleischer, M. (1990) *The Processes of Technological Innovation*, Lexington Books, Lexington, MA.

Tracey, M. (1998) The importance of logistics efficiency to customer service and firm performance, *International Journal of Logistics Management*, Vol. 9, No. 2, pp. 65–82.

Tripsas, M. (1997) Unraveling the process of creative destruction: Complementary assets and incumbent survival in the typesetter industry, *Strategic Management Journal*, Vol. 18, No. S1, pp. 119–142.

Turnbull, P., Oliver, N., and Wilkinson, B. (1992) Buyer-supplier relations in the UK automotive industry: strategic implications of the Japanese manufacturing model, *Strategic Management Journal*, Vol. 13, No. 2, pp. 159–168.

van Hoek, R.I. (1999) From reversed logistics to green supply chains, *Supply Chain Management*, Vol. 4, No. 3, pp. 129–134.

Vokurka, R.J. and Lummus, R.R. (2000) The role of just-in-time in supply chain management, *International Journal of Logistics Management*, Vol. 11, No. 1, pp. 89–98.

Vonderembse, M.A. and Tracey, M. (1999) The impact of supplier selection criteria and supplier involvement on manufacturing performance, *Journal of Supply Chain Management*, Vol. 35, No. 3, pp. 33–39.

Vonderembse, M.A., Tracey, M. Tan, C.L. and Bardi, E.J. (1995) Current purchasing practices and JIT: Some of the effects on inbound logistics, *International Journal of Physical Distribution and Logistics Management*, Vol. 25, No. 3, pp. 33–48.

Voss, C.A. and Robinson, S.J. (1987) Application of Just-in-time manufacturing techniques in the UK, *International Journal of Operations and Production Management*, Vol. 7, No. 4, pp. 46–52.

Wafa, M., Yasin, M. and Swinehart, K. (1996) The impact of supplier proximity on JIT success: An informational perspective, *International Journal of Physical Distribution and Logistics*, Vol. 26, No. 4, pp. 23–34.

Walton, S.V., Handfield, R.B. and Melnyk, S.T. (1998) The green supply chain: Integrating suppliers into environmental management process, *International Journal of Purchasing and Materials Management*, Vol. 34, No. 2, pp. 2–11.

Wang, X. and Cheng, T.C.E. (2007) Machine scheduling with an availability constraint and job delivery coordination, *Naval Research Logistics*, Vol. 54, No. 1, pp. 11–20.

Ward, P. and Zhou, H. (2006) Impact of information technology integration and lean/just-in-time practices and lead-time performance, *Decision Sciences*, Vol. 37, No. 2, pp. 177–203.

Wernerfelt, B. (1984) A resource-based view of firms, *Strategic Management Journal*, Vol. 5, No. 2, pp. 171–180.

White, R.E., Pearson, J.N. and Wilson, J.R. (1999) JIT manufacturing: a survey of implementations in small and large U.S. manufacturers, *Management Science*, Vol. 45, No. 1, pp.1–15.

White, R.E. and Prybutok, V. (2001) The relationship between JIT practices and type of production system, *Omega*, Vol. 29, pp. 113–24.

Whitson, D. (1997) Applying just-in-time systems in healthcare, *IIE Solutions*, Vol. 29, No. 8, pp. 32–37.

Womack, J.P. and Jones, D.T. (2003) *Lean Thinking: Banish Waste and Create Wealth in Your Corporation*, Free Press, New York.

Wu, Y.C. (2003) Lean manufacturing: a perspective of lean suppliers, *International Journal of Operations and Production Management*, Vol. 23, No. 11, pp. 1349–1376.

Wu, Y.N. and Cheng, T.C.E. (2008) The impact of information sharing in a multiple-echelon supply chain, *International Journal of Production Economics*, Vol. 115, pp. 1–11.

Yang, J., Wang, J., Wong, C.W.Y. and Lai, K.H. (2008) Relational stability and alliance performance in supply chain, *Omega*, Vol. 36, No.4, pp. 600–608.

Yasin, M.M., Small, M. and Wafa, M.A. (1997) An empirical investigation of JIT effectiveness: An organizational perspective, *Omega*, Vol. 25, No. 4, pp. 461–471.

Yasin M.M., Small M. and Wafa, M.A. (2003) Organizational modifications to support JIT implementation in manufacturing and service operations, *Omega*, Vol. 31, No. 3, pp. 213–226.

Yasin, M.M and Wafa, M.A. (1996) An empirical examination of factors influencing JIT success, *International Journal of Operations and Production Management*, Vol. 16, No. 1, pp. 19–26.

Yasin, M.M., Wafa, M.A. and Small, M.H. (2004) Benchmarking JIT – an analysis of JIT implementations in the manufacturing service and public sectors, *Benchmarking: An International Journal*, Vol. 11, No. 1, pp. 74–92.

Yeung, A.C.L., Cheng, T.C.E. and Lai, K.H. (2005) An empirical model for managing quality in the electronics industry, *Production and Operations Management*, Vol. 14, No. 2, pp. 189–204.

Yeung, A.C.L., Cheng, T.C.E. and Lai, K.H. (2006) An operational and institutional perspective on total quality management, *Production and Operations Management*, Vol. 15, No. 1, pp. 156–170.

Yeung, A.C.L., Lai, K.H. and Yee, R.W.Y. (2007) Organizational learning, innovativeness and organizational performance: A qualitative study, *International Journal of Production Research*, Vol. 45, No. 11, pp. 2459–2477.

Zaltman, G., Duncan, R. and Hobek, J. (1973) *Innovations and Organizations*, New York, John Wiley & Sons.

Zhu, Z., Meredith, P. and Makboonprasith, S. (1994) Defining critical elements in JIT implementation: A survey, *Industrial Management & Data Systems*, Vol. 96, No. 5, pp. 3–10.

Zhu, Q., Sarkis, J. and Lai, K.H. (2007) Green supply chain implications for closing the loop, *Transportation Research Part E*, Vol. 44, No. 1, pp. 1–18.

Zhu, Q., Sarkis, J. and Lai, K.H. (2008) Confirmation of a measurement model for green supply chain management practices implementation, *International Journal of Production Economics*, Vol. 111. No. 2, pp. 261–273.

Zipkin, P.H. (1991) Does manufacturing need a JIT revolution? *Harvard Business Review*, Vol. 69, No. 1, pp. 40–50.

Zsidisin, G. A. and Hendrick, T. E. (1998) Purchasing's involvement in environmental issues: A multi-country experience, *Industrial Management & Data Systems*, Vol. 98, No. 7, pp. 313–320.

Index

ABC analysis 57, 90
Advanced Planning and Scheduling (APS) 94
adversarial bargaining systems 121
after-sales service 5, 49
Agile Manufacturing (AM) 43
agility 43, 86–87, 103
assurance 54
autonomy 27, 105, 123, 138, 151, 153–154

backhauls 100–101
benchmarking 90, 129, 133–134, 156
bills of materials 88–89
brainstorming 129–130, 132
breakability 97–98
bullwhip effect 83–86, 92, 94, 105

cause and effect diagrams (*see also* Ishikawa diagrams) 129–130, 132–133
cellular manufacturing 13
check sheets 129, 132
Closed-Loop Supply Chain 72
coercion 151, 153–154
collaborative forecasting 93
collaborative planning 92–94
collaborative replenishment 93
communication
    barriers 82
    between the customers and suppliers 70, 108
    between employees 114, 125
    between management and employees 121
    channels 103, 142
    delays 60
    effective 19, 121, 140
    electronic 60–61
    improving 1, 21, 25, 67, 77, 114
    ineffective 115
    infrastructure 75
    instant 139
    inter-organizational 101
    lack of 142
    linkages 76
    open 122, 126, 142
    paper-based 60
    problems 72
    real-time 60–61
    standards 76
    systems 106
    technologies 33–34, 77
    two-way 107, 136
    vertical and horizontal 123
    word of mouth 53
competitive 22–24, 32, 73, 164
    advantage 21, 27, 34, 44–45, 48, 62, 73, 75–76, 87, 96–97, 99, 111–112, 129
    bidding 63
    capabilities 67
    edge 26, 44–45, 69, 107, 112
    environment 39, 69, 145
    implications of quality 112
    marketplace 47
    performance 25
    position 25, 111
    pressure 72, 111
    prices 34
    pricing 68
    priorities for purchasing 67
    requirements 73
    rivals 26
    strategic advantage 30
    strategy 43, 67–68
    strengths 49, 73
    success 157
competitiveness 4, 7, 24, 44–45, 62–63, 128, 139
    international 9

organizational 22
complementary asset 148–149
Container Security initiative (CSI) 102
conformance 23, 68, 71, 116, 151, 157
    quality 29
    quality conformance system 127
continuous improvement 6, 42, 121, 160–161, 164
continuous production 26, 64
control 86–94
    between the buyer and supplier 71
    goods receiving 90
    inventory 34, 39, 84, 90–91, 94, 117
    materials 70, 88
    materials 70, 137
    performance 57, 65
    points 17, 142
    process 107, 142–143
    production 42, 92, 143
    production operations 89
    quality 14–15, 30, 50, 132
    supervisory 126
    systems 10, 75, 92
    total quality 78, 123
    visibility 143
    visual 42
conventional 102–103
    firms 22
    practices 58
    order processing 60
Cooperative Planning, Forecasting and Replenishment (CPFR) 47
Cost/revenue trade-offs 56
credit check 58
cross-docking 91
cross-border supply chains 103
cross-functional
    cooperation 74
    employee training 42
    orientation 122
    programmes 73
    thinking 42
cross-training 143
culture 12, 26, 112, 114, 121–122, 137, 148
    learning 134
    no-blame 136

    of JIT 164
    organizational 12, 19, 114, 135, 141, 162
    social 26
    quality 117
Customs & Trade Partnership against Terrorism (C-TPAT) 102
customer
    complaints 28, 49, 97, 161–163
    demand 3, 9, 11, 21, 24, 39, 81, 88–89, 93, 95, 111
    distribution-intermediary 38
    documentation standards 105
    downstream 111
    expectations 2, 48, 52–55, 67, 72, 74, 111–112, 115–116
    focus 117
    immediate 84
    internal 82
    loyalty 5, 39
    need 23, 32, 38, 43–44, 48–49, 52, 55–56, 59, 77, 88, 92, 114, 116–117, 119, 121, 161, 163
    orientation 11, 55
    outcomes 157
    partnerships 5, 42
    perceptions 49, 51, 53–57, 134, 158
    potential 113, 116
    profile 113
    relation 14
    relationship 17, 47, 74, 145
    relationship management 40, 47
    requirements 6, 20, 34–35, 42, 48–49, 52, 56–57, 70, 87, 94, 96, 111, 117, 121, 123, 155, 161
    response 91
    responsiveness 11, 21, 25, 77, 89
    returns 68, 72
    satisfaction 2, 5, 16–17, 20, 22, 30, 39, 47–49, 52–53, 56–57, 62, 73, 96, 112–113, 117, 123, 161–163
    segment 55
    target 52
    value 40, 49, 69, 111, 113, 128, 157, 159
    want 7, 23, 41, 52–53, 56–57, 91, 117, 163

customer cooperation with environmental concerns 73–74
customer-facing measures 159
customer order 41, 48, 50, 78, 92, 105
   cycle 58–59, 61
   delivery 28
   receipt 28
Customer Packaged Goods (CPG) companies 106
customer service 1, 4–7, 17, 26, 34–39, 41, 44–45, 47–52, 55–58, 69, 72, 76–78, 80, 82, 84, 96, 108
   audit 57
   capabilities 99
   counter 48
   defined 4
   elements 49
   flexibility 99
   hotline 48
   management 40, 56
   management tools 56
   personnel 55
   policy 49, 57
   quality 53
   representative 48
   requirements 6
   strategies 56
   waste 51
customization 82, 119
   mass 88, 92
   process 88
cycle
   lead time 21
   stock 80
   time 42, 58, 60, 106, 155, 158

delivery reliability 45, 64, 68
demand management 40, 79, 94–95
demand forecast 17, 50, 84, 92, 95, 136, 139
demand planning system 95–96
demand pull 11, 19–20, 26, 28, 42
Deming 118
Deming's Plan-Do-Check-Act cycle (*see also* PDCA cycle) 123–124
Design for Environment (DfE, *see also* eco-design) 74

discipline 21, 25, 123, 132
distribution 3, 5, 17–18, 33, 40, 49, 58, 65, 72, 78, 81–82, 144, 161, 163
   capacity utilization 86
   cost 34
   data 131
   elements 96
   functions 82
   intermediary-customer 38
   marketing 38
   networks 34, 99, 104, 143
   nodes 104
   patterns 33
   physical 2, 38–39
   pipeline 107
   routes 64
   services 97
   strategies 33, 93
   system 75, 93, 99
distribution management 143
distribution requirement planning (DRP) 89

eco-design 73–74 (see also Design for Environment, DfE)
ecological efficiency 73
Efficient Customer Response (ECR) 47, 91
Electronic Data Interchange (EDI) 6, 17, 48, 60, 65, 70, 75–78, 90–91, 106, 108, 139, 142, 156
Electronic logistics 77–78
Electronic Product Code (EPC) 105
empathy 54–55, 137
employee
   capability 114
   commitment 125, 163
   education of 18, 31, 138, 142, 147, 162–163
   empowerment 112, 116, 137, 140, 164
   flexibility 25, 31
   Japanese 10
   loyalty 10
   morale 125
   motivation 115, 125
   multi-skilled 14–15, 31
   overtime and undertime 85
   ownership 115, 118, 122, 125

participation 15, 118
readiness and support 142
relations 65
relationships 147
roles of 118
training 22, 30, 42, 116, 126, 137, 147, 163
transfers 152
turnover 152
utilization 116
employee involvement 13, 17, 19, 24, 28, 114–115, 117, 121–123, 125–126, 136–138, 145, 147, 165
defined 137
employee resistance 146–147
enabling technologies 6, 60, 76, 94, 105, 109, 147–148
engineering 1, 13, 67, 89, 107, 118, 123, 137, 142
Enterprise Resources Planning 20, 94
eXtensible Markup Language (see also XML-files) 77

firm infrastructure 5, 44
five S's 132
five whys (5 'whys') 129
flexible manufacturing 89
flip charts 129
focused factory 14–15
forecasting 92–94
    collaborative 93
    demand 84
functional interdependence 69

global competition 1, 128
global supply chain 40, 101–102
Global Positioning System (GPS) 106–107
green purchasing 73–74
Green Supply Chain Management 73

handling
    compliant 7, 28, 49, 116, 163
    defect products 49, 157
    equipment 100
    excessive 28
    facilities 103
    materials 1, 10, 36, 44, 97

order discrepancy 51, 55
returns 49
shipping documents 101
special 100
histogram 129, 131
housekeeping 115, 132
human resources 4–5, 18, 35, 44, 94, 113, 135

idle time 63–64, 99–100, 107
implementation strategy 135, 141–142, 165
importance-performance analysis 57
importance-performance matrix (IPM) 57–58
information quality 55
innovation and knowledge management 128
in-process 14
    inventory 34, 36
institutional isomorphism 150, 152–154
institutional pressures 148–150
internal environmental management 73
International Ship & Port Security Code (ISPS) 102
Internet 76
inventory investment 9, 38, 42, 65
inventory management 1, 4–6, 16, 33, 36–37, 39, 45, 47, 50, 56, 71, 78–79, 81, 91–92, 143, 161
    defined 4, 39
    pitfalls of 82
    tools 88, 90
    wastes in 83, 86, 109
inventory types 79–80
investment recovery 73–74
Ishikawa diagrams (see also cause and effect) 129
ISO 9000 64, 111, 127–128, 142

Japanese
    cultural characteristics 11–12
    management philosophy 31
    work ethic 10, 12, 23
jidoka (see also quality at the source) 14
job rotation 143
Just-in-Time (JIT) 9–12
    benefits of 30

competitiveness 24
defined 81
goals of 21–24
history and development 9
in services 16–18
limitations of 25–27
management approach 1
philosophy 10–11, 16, 60, 87, 157
prerequisites 30–31
principles of 12–16
purchasing 13–15, 65, 81, 142
rationale for 27–30
significance of 5–7
value of 2–4
Just-in-Time customer service 47
Just-in-Time Logistics
elements of 18–21
implementation of 159–165
objective 47, 50–51
significance of 5–6
Just-in-Time manufacturing 13
Just-in-Time order cycle 61–62

Kanban 10–11, 13–14, 19–20, 42, 65, 88, 143
Systems 15, 64

lead time 59, 61, 100, 122
lean inventories 78
lean supply 71
lean supply chain 43
lean thinking 86–87
line balancing 15
localization 82, 143
logistics 1–2, 33–35
activities 1, 23, 33, 35–38, 45, 56, 71, 86, 88, 93, 98, 109, 111–113, 115, 120, 138–139, 141, 144, 155, 157, 161
competitiveness 44
defined 34
elements 39
goals of 43
history and development 33
inbound 4–5, 38, 44–45, 75
mix 36–37, 45, 47, 56, 98, 108–109
network 18–23, 32

outbound 4–5, 38–39, 44–45, 75
scope of 35
system 35
value of 4–5
Logistics Information System (LIS) 6, 78
logistics service quality 5, 55, 134
lot sizes 15, 24, 48, 78

management commitment 65, 117, 159–160
manufacturing 1, 5, 9–11, 13–16, 24, 27–28, 40, 89, 93–94, 120
manufacturing cells 15, 42
manufacturing flow management 40
Manufacturing Resources Planning (MRP II) 89
marketing 1, 4–5, 17–18, 38, 44, 72, 79, 81, 89, 123, 164
materials handling 1, 10, 97
Materials Requirement Planning (MRP I) 20, 88
measurement systems 153, 155
mimesis 151–154
mix flexibility 68
mixed model scheduling 14
mode shifting 103
modification flexibility 68
motion 6, 34
    the waste of 28
multifunctional workers 13, 19–20, 31
multi-echelon distribution 143
multi-skilled employees 14–15, 18
multi-skilled workforce 14
multiple-source suppliers 66
multiple sourcing 63

nominal group technique (NGT) 129
    nonconformance 25, 68, 116, 121
non-value-added (NVA) activities 10, 121, 127

Ohno, Taiichi 9
oil embargo 3, 9
order accuracy 55
order batching 84
order condition 55
order discrepancy handling 55

order fulfillment 40, 94
order processing 1, 4–6, 34, 36–39, 45, 47, 49, 56, 58–61, 75, 84, 97, 108–109, 163
order quality 55
order release quantities 55
ordering procedures 55
Organization Theory 19
organizational culture 12, 19, 114, 135, 141, 162
organizational factors 135–136, 165
organizational flexibility 31
organizational innovativeness 148
overproduction 11, 28

packaging 1, 12, 28, 33, 35–36, 49–50, 63, 72, 97–100
Pareto analysis 129
Pareto chart 131
Pareto principle (Pareto's Law, 80-20 rule) 87, 90
PDCA cycle (*see also* Deming's Plan-Do-Check-Act cycle) 123–124
people involvement 18–19, 22, 32
performance measurement 2, 7, 82, 153, 155, 157–158, 162, 165
performance measures 37, 57, 82, 158–159
perishability 97–98
Persian Gulf War 33
Personal contact quality 55
pilot programme 124
pilot projects 145–146
plant layout 24, 29
point-of-sale (POS) 17, 78
    system 94
poka-yoke error prevention 14
postponement 88, 92
preventive maintenance 13–15, 24, 164
preventive measures 103, 116
proactive approach 147, 163
product-process design 82
product pricing 85
product rationing 85
product relationship management 40
product stewardship 74
production-oriented 2, 89
process improvement 32, 50, 88, 123

processing 11, 15
processing time 28
product defect 72, 158
product durability 68
product innovation 67–69, 90
product reliability 68
production control system 10, 143
production management 13, 38, 143
production practices 25
pull system 11, 20–21, 27
purchasing 1, 16, 21, 25, 30, 33–34, 36–38, 62–68, 75, 80–81, 90
purchasing management 17, 38, 143
push system 20

quality assurance 67, 120, 127
quality at the source 14, 20, 42
quality awareness 111, 162
quality circles 13–16, 19, 24, 30, 121, 125–127, 136
    defined 124
quality control 14–15, 30, 50, 123, 132
quality cost 23, 113
quality management 7, 20, 65, 109, 111–113, 116, 134, 143
    supplier 70
    system 111, 120
    tools 129
quality service 52, 120, 162
quality gaps 53
queuing time 28
quick response 16, 91, 123

Radio Frequency Identification (RFID) 17, 106–107, 148–150
reliability 54
resistance to change 18, 114, 126, 147
resource allocation 94–95
Resource-Based View (RBV) of the firm 129
resource management 94–95
resource optimization 95
Responsive Supply Chain (RSC) 43
responsiveness 3, 11, 40, 54, 56, 77, 120, 158–159
return management 40, 71
Return on Investment (ROI) 44

reverse logistics 71–72, 74

safety stocks 6, 64, 78, 80, 84, 86, 95–96
sanitize (*see also* Seiso) 132
schedule stability 24
Seiketsu (*see also* standardize) 132
Seiri (*see also* structurize) 132
Seiso (*see also* sanitize) 132
Seiton (*see also* systematize) 132
self-discipline (*see also* Shitsuke) 132
self-inspection 19–20, 116
sell one-make one (SOMO) 11
service-oriented economy 2
service discrepancies 51
service improvement 2, 6–7, 14, 36, 40, 42, 88, 100, 105, 109, 111, 113, 135–138, 144, 165
service quality 5, 26, 51–55, 111–112, 120, 134, 158, 162
SERVQUAL 54–55
set-up time 13–15, 21–22, 24, 29–31, 63, 157–158
seven rights (7Rs) 4, 6, 20, 43, 109, 111, 165
seven wastes 28
shape 50, 59, 97–98, 157
Shitsuke (*see also* self-discipline) 132
simplifying 9, 14–15
single sourcing 70, 140
small batch 42, 65, 87
standardize (*see also* Seiketsu) 132
standardizing 154
standardization 14–15, 132
statistical process control 143
stock
  cycle 80
  dead 80–81
  excessive 81
  idle in-transit 81
  in-transit 80
  out-of 81
  safety 80
  seasonal 80
  speculative 80
strategic control points 142
structurize (*see also* Seiri) 132
substitutability 97–98

suggestion systems 122
supplier relationships 63, 69–71, 79
supplier relationship management 40, 103, 156
supplier selection 65–67
Supply Chain Management 11, 40, 42
sustainable
  competitive advantage 34, 44–45, 112, 129
  competitive edge 44–45
  growth 4
systematize (see also Seiton) 132
systems 20
systems perspective 113

tangibles 54
technological capabilities 68
technological opportunism 148
technology 4–5, 17, 44, 50, 60, 70, 73, 75, 94, 105–107, 147–149
  electronic ordering 85
  group 14–15
  information 43, 60, 77, 122, 135, 139, 141, 149–150, 162
  sharing 68
technology-response capability 148
technology-sensing capability 148
timeliness 3, 7, 23, 50, 55, 115, 120, 155
tooling 15
top management support 136–137, 142
total business concept 155
Total Employee Involvement (TEI) 13, 118, 122
Total logistics management costs 159
total JIT system 10
total optimization 22
total preventive maintenance 14
total productive maintenance 15, 42
Total Quality Control (TQC) 78, 123
Total Quality Management 111
Toyota Production System 87
traditional manufacturing 138
transportation 4, 36–39, 96–98
  information 100
  links 102
  modes 49, 59, 97–98
  nodes 102

security 101–102
services 100–101
system 96–97
wastes 99
transportation management 1, 39, 47, 50, 96–97, 99
    system 105
    tools 104
transportation network 104

uniform plant loading 14–15
union involvement 121
utility
    form 50
    place 50
    possession 51
    time 50

value-added activities 81, 159
value-added concept 28, 116
Value-Added Network (VAN) 6, 76
value-added productivity 159
value-added services 47
value-weight ratio 98
value chain 4–5, 21, 32, 44–45, 78
Vendor Managed Inventory (VMI) 47, 90

visibility 42, 47, 66, 76, 86, 107, 114–115, 134, 143
visible signals 12
visual control 42
volume 97
volume flexibility 68

warehouse management system 105
warehousing 1, 26, 34, 36–38, 44, 48, 94, 96–97, 101, 120, 143
waste
    defined 3, 5
    reduction 3, 5–7, 29, 32, 47, 52, 65, 163–164
weight 97–98
weight-bulk ratio 98
work-in-process (WIP) 6, 10–11, 17, 19, 24, 30, 50, 79–81, 114
work ethnic 9–10, 12, 23
World Wide Web 107
XML-files (*see also* eXtensible Markup Language) 77

zero-defect 17, 161
zero-inventories 10, 78, 81

Name Index

3M 37
7-Eleven 4, 41

Ahire, S.L. 112
Aichlmayr, M. 26
Alexander the Great 33
Alles, M. 24, 83
Alternburg, K. 107
Amazon.com 49
Ansari, A. 99
Antonakis, J. 29
Asubonteng, P. 52
Awstorm, P. 13

BAAN 139
Bagchi, P. 90, 98
Baker, W.E. 128
Balakrishnan, R. 24,
Ballou, R.H. 4, 34–36, 38–39, 48–49, 92
Barlow, G.L. 17
Barnerjee, A. 65
Beer, M. (2003) 136
Benito, J. 65
Benson, R.J. 16
Benton, W.C. 71
Berman, S.L. 18
Bhargava, P. 87
Bichou, K. 103
Billesbach, T. 6
Billington, C. 81, 140
Bookbinder, J. H. 61
Bose Corporation 10
Bowen, F.E. 73
Bowersox, D. 34, 76, 94
Boyle, M. 148,
Brinker, B. 116
British Airways 16

Callen, J.L. 14, 25
Campbell Soup 4
Canel, C. 17, 24
Cannon, J.P. 71
Carter, J.R. 73
Carter, P.L. 24
Chakravarthy, B. 129
Chandy, R.K. 148
Chapman, S.N. 24
Chappell, G. 106, 149
Chase, R.B. 3, 15, 88,
Cheng, T.C.E. 9, 24–25, 40, 57, 78, 86, 102, 107, 111–112
Chopra, S. 52
Christopher, M. 33, 40, 44, 86, 87
Closs, D.J. 71
Clouse, V. 76, 101, 104
Coca-Cola 4
Cohen, W. M. 128, 149
Cooper, M.C. 71
COSO 108
the Council of Logistics Management (*see also* the Council of Supply Chain Management Professionals) 34
the Council of Supply Chain Management Professionals (*see also* the Council of Logistics Management) 34
Coyle, J. 35, 90, 94
Croxton, K.L. 40
Cusumano, M.A. 70

Davy, J.A. 3, 14,
De Toni, A. 70
de Treville, S. 29
Dilts, D. M. 61
DiMaggio, P.J. 150
Domino Pizza 52
Dong, Y. 25

Dow Jones & CO., Inc 35
Dowlatshahi, S. 72
Dreyfus, L.P. 14
Droge, C. 25, 81, 137
Drucker, P.F. 12
Dwyer, F.R. 71

Ebrahimpour, M. 14
Epps, R. 81
Evergreen 108

Federal Express 17
Fielder, K. 13
Finkenzeller, K. 107
Fitzsimmons, J.A. 14
Flapper, S.D.P. 72
Fleischer, M. 149
Fliedner, G. 111
Ford Motor Company 26
Frazelle, E. 104
Frazier, G.L. 67
Fullerton, R. 3, 11, 14, 25
Fynes, B. 71

Garcia-Dastugue, S.J. 87
Garvin, D.A. 111–112
Gélinas, R. 141, 143
Gentry, J. 99
Germain, R. 14, , 17, 21, 25–26, 81, 137
Giunipero, L. 66
Goldsby, T.J. 87
Gonzalez-Benito, J. 64
Gourdin, K. 33–34
Graham, I. 13
Grapentine, T. 53
Green, K.W. 16–17
Gunasekaran, A. 25, 43, 65
Gupta, Y. 76, 101, 104

Hall, E. Jr. 26
Hall, R.W. 31
Hallihan, A. 13, 28
Ham C. 53
Hamel, G. 73
Hampson, J. 71

Hartley, J.L. 65
Hay, E.J. 20
Hayen, R. 6
Heiko, L. 11
Hendrick, T.E. 74
Herman Miller 37
Heskett, J. 36
Hewlett Packard 41
Hill, T. 16
Homburg, C. 71
The Home Depot 106
Hopp, W.J. 29
Houlihan, J.B. 40
Hult, G.T.M. 129
Hurley, S. 16
Huson, M. 14
Hwang, Y.D. 42

Im, J.H. 13
Inman, R.A. 14, 16, 24,
The International Maritime Organization (IMO) 102

James, J.C. 57
Jaworski, B.J. 56
Jayaraman, V. 73
Jayaram, J. 70, 137
Johnson, J. 33
Jones, D.T. 75, 87

Kaneko, J. 2
Kannan, V.R. 41
Kaplan, R.S. 129, 157
Karlson, C. 13
Kay, J.M. 87
Kaynak, H. 65
Ketchen, D.J. 129
Kim, S. 65
King, J.L. 149
Kinney, M.R. 3, 11, 24
Kleenex 4
Kohli, A.K. 56
Koufteros, X.A. 67, 122–123
Krajewski, L.J. 15
Krause, D.R. 67
Krikke, H. 74

Lai, K.H. 5, 40, 56–57, 64, 70–71, 75, 77, 85, 102, 107–108, 111–112, 139, 148, 150, 158
Lambert, D.M. 33, 35, 37, 40
Lamming, R. 71
Lance Dixon 10
Lee, H.L. 41, 79, 81, 83–84, 86, 88, 101, 140
Lee, S.M. 13
Levesque T. 53
Levinthal, D.A. 128, 149
Li, J. 92
Lin Y.C. 42
Lines, R. 19
Lowes 52
Lu, C.S. 101, 108
Lummus, R.R. 12, 25, 42
Lun, Y. H. V. 96, 103
Lyu, J. 42

Maersk 108
Maloni, M. 71,
Martilla, J.A. 57
Massey, A.P. 129
Matsui, Y. 25
McDonald's 16, 35
McDougall G. 53
McEvily, S.K. 129
McGillivray, G. 26
McWatters, C. 3, 11, 14, 25
Mehra, S. 14, 24
Meindl, P.52
Mentzer, J.T. 55, 139
Mia, L. 13
Modarress, B. 99
Mollenkopf, D.A. 71
MSC 108

Nagel, R.N. 87
Nanda, D. 14
Narasimhan, R. 73
Nassimbeni, G. 70
Ngai, E.W.T. 76, 78, 107
Nojiri, W. 4
Norton, D.P. 129, 157

Oakland, J.S. 112
Ohno, T. 9,

O'Neal, C. 66
OOCL 108
Oracle 139
Orton, C. 10

Pagell, M. 73
Parasuraman, A. 53–54
PeopleSoft 139
Pine, J.B. 67
Plenert, G. 20
Podolsky, S. 9, 24–25
Polito, T. 141
Porter, M. 4, 21, 44, 75
Powell, W.W. 150
Power, D. 19
Prahalad, C.K. 73
Prince, J. 87
Proctor & Gamble 16, 106

Quaker Oats 37

Ritzman, L.P. 15
Robinson, S.J. 13
Rogers, E.M. 147, 149
Russell, D.M. 103

Sakakibara, S. 14
Salaheldin, S. 138
Saldanha, J.P. 103
SAP 139
Scannell, T.V. 70
Schniederjans, M.J. 15
Schonberger, R.J. 9, 14, 78, 81
Sears 35
Seuring, S. 73
Sheffi, Y. 101
Simchi-Levi, D. 41
Singh, N. 75
Sinkula, J.M. 128
Sitkin, S.B. 112
Sohal, A. 13, 19
Sowinski, L. 10
Spearman, M.L. 29
Srinivasan, R. 148
Stank, T.P. 5, 70
Stanley, L.L. 70
Stock, J.R. 35, 43

Takeishi, A. 70
Tan, K.C. 41
Teo, T. S. H. 60
Teece, D.J. 149
Tellis, G.J. 148
Thierry, M. 74
Thompson, V.A. 148
Tibben-Lembke, R.S. 74
Toffel, M. 74
Tornatzky, L.G. 149
Towill, D.R. 87
Toyota 2–3, 9, 18, 23
Toyota Motor Corporation 88
Tracey, M. 5, 70
Tripsas, M. 149
Turnbull, P. 13

Unilever 106

van Hoek, R.I. 74
Vickery, S.K. 137
Voss, C. 13, 71
Vokurka, R.J. 25, 42, 111
Vonderembse, M.A. 70, 99
Voss, C.A. 13, 71

Wafa, M. 18, 21, 23, 60
Wal-Mart 4, 16–17, 52, 106, 148–150

Walton, S.V. 73–74
Wang, X. 40
Ward, P. 25
Watson, K. 141
Wempe, W.F. 3, 11, 24
Wernerfelt, B. 129
Whang, S. 101
Whirlpool Corporation 37
White, R.E. 12–14
Whitson, D. 17
Whybark 16
Wisner, J.D. 70
Womack, J.P. 75, 87
Wong, C.W.Y. 107
World Customs Organization (WCO) 102
Wu, Y.C. 9
Wu, Y.N. 78, 86

Yang, J. 71
Yasin, M.M. 17–19, 21, 23, 25, 29–30
Yeung, A.C.L. 112, 116, 128

Zaltman, G. 148
Zhou, H. 25
Zhu, Q. 73
Zhu, Z. 18
Zipkin, P.H. 12
Zsidisin, G.A. 74